Better Homes and Gardens®

step-by-step

vegetables

Rita Buchanan

Better Homes and Gardens® Books
Des Moines, Iowa

Better Homes and Gardens® Books
An imprint of Meredith® Books

Step-by-Step Vegetables
Senior Editor: Marsha Jahns
Production Manager: Douglas Johnston

Vice President and Editorial Director: Elizabeth P. Rice
Executive Editor: Kay Sanders
Art Director: Ernest Shelton
Managing Editor: Christopher Cavanaugh

President, Book Group: Joseph J. Ward
Vice President, Retail Marketing: Jamie L. Martin
Vice President, Direct Marketing: Timothy Jarrell

Meredith Corporation
Chairman of the Executive Committee: E. T. Meredith III
Chairman of the Board and Chief Executive Officer:
    Jack D. Rehm
President and Chief Operating Officer: William T. Kerr

Produced by ROUNDTABLE PRESS, INC.
Directors: Susan E. Meyer, Marsha Melnick
Executive Editor: Amy T. Jonak
Editorial Director: Anne Halpin
Senior Editor: Jane Mintzer Hoffman
Design: Brian Sisco, Susan Evans, Sisco & Evans, New York
Photo Editor: Marisa Bulzone
Assistant Photo Editor: Carol Sattler
Encyclopedia Editor: Henry W. Art and Storey
    Communications, Inc., Pownal, Vermont
Horticultural Consultant: Christine M. Douglas
Copy Editor: Sue Heinemann
Proofreader: Cathy Peck
Assistant Editor: Alexis Wilson
Step-by-Step Photography: Derek Fell
Garden Plans: Elayne Sears and Storey Communications, Inc.

All of us at Meredith® Books are dedicated to providing you with the information and ideas you need for successful gardening. We guarantee your satisfaction with this book for as long as you own it. If you have any questions, comments, or suggestions, please write to us at:

Meredith® Books, *Garden Books*
Editorial Department, RW206
1716 Locust St.
Des Moines, IA 50309–3023

STEP-BY-STEP

Vegetables

A Passion for Vegetables

growing vegetables is one of the most popular forms of gardening. Just bite into a plump, juicy vine-ripened tomato. You can't buy flavor like that in any store. And tomatoes are just the top of the list. All kinds of vegetables, even humble staples like onions and potatoes, taste much better when you grow them yourself. • Flavor and freshness are two reasons for growing vegetables. Increased variety is another benefit. Instead of the one or two kinds you can find at a store, any seed catalogue lists dozens of unique types of tomatoes, corn, and other crops that you can grow. Part of the fun is comparing them and choosing your own favorites. • Growing vegetables is very gratifying. The plants develop quickly, and they respond generously to good care. Only two or three months after planting, you'll be picking all you can eat, with extra to share. No other form of gardening offers such prompt and bountiful rewards.

Seed Savers Exchange

The Seed Savers Exchange is a network of gardeners who are committed to preserving heirloom vegetables. They do not sell seeds, but they welcome new members who want to participate by sharing seeds. They publish an annual yearbook and sponsor the large Heritage Farm, where many of the varieties are grown each year. For more information, write to Seed Savers Exchange, Rural Route 3, Box 239, Decorah, IA 52101.

*T*raditionally, growing vegetables was a matter of feeding the family, and it was serious hard work. It took a lot of space, time, and effort to plant, tend, harvest, and preserve a year's supply of potatoes, beans, corn, tomatoes, onions, carrots, cabbage, squash, and other crops. Today, few Americans rely on their gardens for subsistence. Now most of us grow vegetables because we *want* to, not because we *have* to. This change in circumstances and attitudes has led to several new developments in vegetable gardening.

One of the most significant trends in vegetable gardening is the movement toward organic methods. Only 10 years ago, most people were still using synthetic fertilizers and pesticides in their vegetable gardens, and organic gardening was considered a fringe movement. Now it's very popular, because it addresses widespread concerns about safety and the environment. Composting has become as popular as other forms of recycling, and gardeners have learned that improving the soil with compost and other organic amendments makes it easier to grow vigorous, high-yielding plants. More and more people have turned to nontoxic pest controls so that they can grow pure, wholesome vegetables without dangerous residues. This book includes many organic ideas and techniques that will help you build healthy soil and grow healthy plants.

Another contemporary trend is a growing concern about genetic diversity. To vegetable gardeners, this means preserving the thousands of heirloom varieties that have been passed down from generation to generation. Some of these varieties have outstanding flavor or beauty. Others are especially well adapted to a particular climate, or unusually tolerant of stressful conditions such as cool temperatures or drought. You may find such a treasure in your own garden or a neighbor's—a tomato that makes extra-thick sauce, or a pole bean that keeps producing all summer. Although heirloom vegetables have many valuable traits, most were on the verge of extinction in the late 1970s, replaced by a narrow selection of modern hybrids. But now many of the heirlooms have been rescued and some have become quite popular, thanks to the commitment of a few key individuals—especially Kent Whealey, founder of the Seed Savers Exchange—and a community of dedicated gardeners who never stopped saving and sharing their own special seeds.

Americans have always enjoyed regional food specialties. Now the proliferation of ethnic restaurants has introduced us to the food of other cultures, including many interesting new vegetables. Several ethnic specialties that you might discover in a restaurant can also be grown at home. If you plan ahead and order the seeds from a catalogue, you can plant Oriental favorites like snow peas, bok choy, and edible chrysanthemum, or European specialties like chicory, filet beans, and radicchio. You can also grow tomatillos and chili peppers from Mexico, or blue corn and tepary beans from Native Americans in the Southwest. Trying new kinds of vegetables is as much fun for a gardener as it is for a cook. You'll enjoy watching how plants from other places grow and perform in your climate.

This classic vegetable garden is designed for maximum yields. Full sun, good soil, and careful weeding allow the plants to produce loads of vegetables to eat fresh, to freeze, or to can.

At the Seed Savers Exchange garden, screen cages enclose blooming pepper plants to prevent cross-pollination and keep the strains pure.

These are some of the potato varieties being preserved through the efforts of the Seed Savers Exchange and its members.

An array of heirloom sweet peppers includes bell peppers and narrow fruits in red, green, orange, yellow, and purple.

Preserving heirloom varieties ensures that a diverse genetic base remains available to researchers breeding new plant varieties.

One of these heirloom tomato varieties may possess a gene that could allow breeders to solve a future disease or insect problem.

The Seed Savers Exchange eggplant collection shows a surprising range of sizes, colors, and shapes.

Varieties for Baby Vegetables

There are two kinds of baby, or miniature, vegetables. Some are small because they are picked while they are still immature. For example, cucumbers and summer squashes are indescribably tender if picked when just a few inches long. All kinds of potatoes make "new potatoes" if harvested when they reach the size of table tennis balls. The baby corn used in Oriental stir-fries is a type of popcorn picked as soon as the ears begin to form.

Other baby vegetables are true dwarfs that never get big, even at maturity. Thumbelina carrots, Tom Thumb lettuce, Sweet 100 tomatoes, Baby Bell eggplants, and White Pearl onions are perfect miniatures, cute as can be, with all the flavor of their full-size cousins.

One of the greatest rewards of vegetable gardening is the ability to harvest a cornucopia of produce at the peak of perfection.

Discriminating cooks know that even familiar vegetables such as potatoes come in common and gourmet varieties. What distinguishes a gourmet vegetable? In many cases, such as Brandywine tomatoes and Delicata squashes, it's the flavor. Golden beets and Yukon Gold potatoes are valued for their warm gold color. Thumbelina carrots and Lemon cucumbers score points for their dainty size. Sometimes gourmet varieties aren't as productive as more common garden varieties, but they're worth the space if you want to grow and serve something really special.

Making the most of limited garden space is a common challenge today. Fortunately, you don't need a place in the country in order to grow some vegetables—even a city balcony can accommodate a window box of salad greens and herbs and a pot large enough for a cherry tomato or pepper plant. When garden space is limited, you can increase your harvest by planting compact varieties. Check your seed catalogues for bush (rather than vining) forms of cucumbers, squash, and melons; determinate (meaning they reach a certain size and stop growing) tomato plants; and early-season sweet corn. All of these yield normal produce on smaller-than-normal plants, so you can space them closer together and fit more plants into the garden.

Another solution to the problem of limited space is edible landscaping, one of the most creative new ideas in gardening. In edible landscapes, vegetables, herbs, and fruits are liberated from the kitchen garden and invited to mingle with flowers and shrubs all around the house, even in the front yard. The result can be as attractive as it is productive. Think about delicate rosettes of leafy lettuce, tufts of ferny carrot tops, shiny ripe peppers or glossy eggplants, okra blossoms as round and yellow as the full moon, or a feathery cloud of asparagus foliage. Many vegetables are very ornamental, and if you grow your own, you can enjoy them twice—in the garden and on the table.

Baby vegetables are delicate in flavor and texture. Clockwise from top left are Cherry Wonder tomatoes, small-rooted Thumbelina carrots, Bambino eggplants (which don't have to be salted and drained like larger eggplants to remove bitterness), baby cauliflower (available in green-headed varieties as well as the traditional white), and baby summer squash. Harvest the baby squash when it is just a few inches long, even if the flower is still attached.

CHAPTER 1

Designing Your Vegetable Garden

*d*esigning a garden doesn't happen in a flash. It's a step-by-step process of asking yourself questions and looking for answers. The first step is finding a suitable site, determining how big your garden will be, and outlining its shape. • Then you have to decide how to divide the space. Consider if the plants will be grouped in rows, blocks, or beds. Where will the paths go, and how wide will they be? Make a list of the vegetables you especially want to grow, then determine how much of each you should plant. Can you sketch a planting diagram that includes all your favorites and allows enough room for everything to grow? • Your design won't be perfect at first, but it's a way to get started. As you continue to work in the garden, you'll ask new questions and think of new solutions, gradually refining the design from year to year, until finally it's just right—efficient, productive, and beautiful.

Choosing the Site

When you're planning to make a vegetable garden, start by walking around your property and evaluating potential sites. Plenty of sun is the first requirement for growing vegetables, and should be your first consideration in choosing a site. In most parts of the country, full sun all day is best. Across the southern states from California to Texas to Florida, the ideal site has full sun from dawn to noon, with light shade from the afternoon heat. Some vegetables require less sun than others, and there are strategies for coping with shady sites (see page 55), but if you have a choice, pick the sunniest possible site for your vegetable garden. If you're considering a site with trees or buildings nearby, remember that shadow patterns change with the seasons. A site that's sunny in the summer may be shady by fall. Part of evaluating a site is watching where the sun reaches at different times of day and different times of year.

Vegetable plants need plenty of water, so try to choose a site within reach of a faucet. Even in rainy climates, you'll occasionally need to use a watering can or hose and sprinkler. In dry climates, you may want to install an automatic watering system. Either way, it's convenient if there's a tap nearby.

It's uncommon to find a plot that already has soil good enough for growing vegetables. More typically, soil is compacted, infertile, and low in organic matter. Don't worry. It's not too difficult, and it's very rewarding, to transform a barren lot into a bountiful garden by improving the soil.

Watch out, though, for two kinds of problems. Avoid low-lying sites where water collects and stands

An open site in full sun where the land gently slopes is an ideal place for a vegetable garden.

Make the most of the space available to you. Here, the front yard has become a bountiful, beautiful garden.

TIMESAVING TIP

A vegetable garden produces plenty of material to compost—tomato and bean vines, carrot tops, cornstalks, broccoli stems, even weeds. You can save a lot of trips back and forth by locating a compost barrel, bin, or pile in or close to the garden.

on the surface after heavy rains or snowmelt. Even if you were to make raised beds for the plants, walking on muddy paths is no fun. Also avoid sites adjacent to busy thoroughfares or old painted wooden houses, where the soil may be contaminated by lead that has accumulated over the decades from vehicle emissions or flaking paint.

Take a look at the plants that are already growing on your potential garden site. If they look vigorous

and healthy, chances are the site will be suitable for vegetables, too. However, you'll have to remove those existing plants—tops and roots—before you can start anew, and some weeds are hard to eliminate. If you don't recognize them, ask a gardening neighbor or friend to identify the plants on your site and point out potential troublemakers.

Choosing the Site CONTINUED

Ask neighbors about other troublemakers, too, such as deer, rabbits, woodchucks, gophers, crows, loose dogs, and mischievous children. You may have to put a fence or barrier around the garden to safeguard your harvest.

You'll need a place to store the various tools and equipment you use in the garden, and it's handy if this is nearby, whether in the garage or basement or in a special garden shed. Avid vegetable gardeners tend to accumulate quite an inventory of tools and other equipment—a digging fork or spade, hoe, rake, cultivator, hand tools, buckets and baskets, hoses, a rotary tiller and perhaps a chipper or shredder, a garden cart or wheelbarrow, tomato cages, stakes and trellises, row covers, nursery flats and pots, and so

on. Some implements are used daily, others just occasionally or seasonally, but they all serve a purpose—if you can find them when you need them.

Finally, think about appearances. If you keep it neat and tidy, even during the dormant season, a vegetable garden will make you smile every time you look at it. It's fun to have a garden like that right next to the house, or even in the front yard. But you won't enjoy looking at your garden if you let it get weedy or don't clean up in the fall. If you don't have the time or aptitude to keep up with maintenance, consider hiding your vegetable garden in the backyard or behind a hedge or fence.

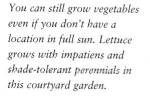

You can still grow vegetables even if you don't have a location in full sun. Lettuce grows with impatiens and shade-tolerant perennials in this courtyard garden.

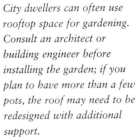

City dwellers can often use rooftop space for gardening. Consult an architect or building engineer before installing the garden; if you plan to have more than a few pots, the roof may need to be redesigned with additional support.

Lettuce grows in partial shade in a space-efficient, narrow raised planter in a city garden.

The sunny, narrow strip of land alongside this driveway makes a productive garden that is easily accessible by wheelchair.

Shapes for Gardens

*T*here are several reasons why a simple rectangle is the traditional shape for a vegetable garden, especially for large gardens of 1,000 square feet or more. If you're going to carve a garden plot out of an open lawn or field, you should consider these advantages. A rectangle is a convenient shape to work if you use a tractor or rotary tiller to plow or till the soil each year. It's easy to calculate a rectangle's area in square feet if you want to apply measured amounts of lime or fertilizers such as rock phosphate, greensand, or 10-10-10. Anyone with a measuring tape and ruler can draw a map of a rectangle and divide it into paths and planting areas. Most garden supplies and accessories such as plastic mulch, floating row covers, portable cold frames, trellises, and irrigation systems are designed to fit into rectangular spaces.

Rectangular gardens are convenient, but some properties can't accommodate a rectangle, and some people think rectangles are boring. Fortunately, plants don't care. You can have a bountiful garden in any shape you choose, regular or irregular. Dig up an awkward space between two paved areas. Make a circle or an oval in the lawn. Trace an undulating curve along the front of a border that backs up to a wall or fence. Terrace a hillside into a series of beds that follow the contour lines.

The main challenge in designing a nonrectangular vegetable garden is providing easy access to all parts of it. It's hard to reach more than 2 feet into a bed to work the soil, sow seeds or set out transplants, spread mulch, pull weeds, check for insect pests, or gather the harvest. You'll probably need to divide the space into a series of smaller beds connected by a network of paths. It's best to plan this in advance and choose specific pathways. Then make a habit of walking only in the paths, to avoid compacting the soil in the beds.

If your climate has cool springs, temperate summers, and plenty of rainfall, consider making one or more raised beds. By raising the soil level 6 to 12 inches or more above the surrounding grade, you create a bed that thaws earlier in the spring, gets warmer in the summer, and drains faster after a rain. Raising the soil level also makes it easier to reach out and care for the plants—you don't have to bend over so far. In general, raised beds should be 3 to 4 feet wide, as long as you choose, and far enough apart for you to walk and carry tools between them with ease.

You can make a raised bed by simply mounding up the soil, but the sloping sides erode easily, and rainfall or irrigation water tends to run off too fast. To stabilize the sides and encourage water to soak in, edge the bed with a frame of landscape timbers, old railroad ties, cement blocks, or stones. Loosen the soil in the bottom of the bed, and fill it with a mix of topsoil (borrowed from the adjacent paths or from other parts of the garden) and organic matter.

Instead of raised beds, try sunken beds in desert climates where the summers are very hot and dry, or anywhere the soil is sandy and drains too fast. Planting in a sunken bed lets the plant roots go deeper in the ground, where the soil is cooler and moister. Sunken beds use water efficiently, because all the water you apply to a bed soaks down into the ground, with no wasteful runoff. To make a sunken bed first dig down and remove the soil to a depth of 1½ to 2 feet. Set aside the topsoil, and use the subsoil to sculpt a berm (or ledge) around the edge of the bed. Then mix the topsoil with organic matter and use it to refill the hole.

Vegetable gardens don't have to be rectangular. This one is circular, with brick paths dividing the circle into pie-shaped wedges.

Neat triangular beds with carefully clipped edgings and paved paths make the vegetable garden above a formal affair.

Creative design can make the most of a small space. With this elegant raised-bed system, the plants are easily accessible, and the pebbled pathways need no weeding. The triangular bed adds interest to the design.

Planting Patterns

One way to gain easy access to plants is to create a grid-like effect with a series of small rectangular beds surrounded by paths. These beds are planted in rows.

Within a garden or bed, you can arrange the plants in different patterns, such as rows or blocks. Each planting pattern has its pros and cons, and is suitable for particular kinds of plants. Choosing different arrangements for different plants makes efficient use of space, simplifies maintenance, and makes a garden look interesting and attractive.

In single rows, individual plants are aligned about 3 to 4 feet apart. Spacing within the row depends on the kind of plant, ranging from 3 to 4 inches apart for small plants like beets or carrots to 1 or 2 feet apart for large plants like tomatoes or okra. Most large-scale vegetable growers plant all their crops in single rows. One advantage is that they can cultivate and weed between the rows with a tractor or rotary tiller, instead of by hand. Another reason is that harvest is simple and efficient—nothing gets overlooked when you go back and forth across a field, picking one row at a time.

Planting in single rows is practical in home gardens for vegetables that need to be picked at frequent intervals over an extended season, such as green beans, Southern peas, tomatoes, or okra; or for plants that require repeated maintenance, such as leeks, celery, or

The traditional single-row method of planting allows you to water, fertilize, weed, and harvest without walking on the planting areas.

potatoes, which need to have soil hilled up around the base as they grow. Otherwise, planting in single rows wastes precious space in small gardens.

You can save a lot of space by planting in wide rows or bands. For example, instead of sowing pea seeds in a single row, make a shallow trench 4 to 6 inches wide and scatter the seeds across its width. This makes such a bushy hedge of pea vines that you can pick quarts of pods from just one short, wide row. Lettuce, spinach, and other greens, plus small root crops such as radishes, carrots, and beets, also do well in wide rows, since the individual plants don't

need much space or much attention. Sow the seeds in several parallel grooves about 4 inches apart. The only problem with wide rows is that you can't hoe or till between the plants to control weeds—you have to pull them by hand. There won't be many weeds, however, if you spread plenty of mulch between the plants while they are still small.

Blocks are groups of plants with equal spacing in all directions. For example, you could make a block of garlic plants spaced 6 inches apart, one of popcorn spaced 1½ feet apart, or one of sweet potatoes spaced 4 feet apart. This is an efficient way to pack as many

Planting Patterns CONTINUED

TIMESAVING TIP

For a care-free crop of onions, plant the sets in a wide row or block and cover immediately with a 3-inch layer of mulch. Onions will grow right up through a mulch, but most weeds will not.

plants as possible into a given area, but it doesn't leave any space for you to walk in. It works best when you can sow or set out the plants, spread mulch once to control weeds, then come back at the end of the season to harvest the whole crop at once.

Corn and various vining plants such as squash, cucumbers, and melons grow well if planted in hills. A hill is a group of three or more plants, traditionally (but not necessarily) planted atop a small mound of soil. A major advantage of planting in hills is that you need to loosen and improve the soil only right around the planting hole. You don't need to cultivate the spaces between hills, which can be 3 to 6 feet apart.

Intensive planting is a strategy for using space efficiently by combining plants with different growing seasons and habits. For example, you might use the wide spaces between single rows of summer-ripening tomatoes to grow early crops of kohlrabi, cauliflower, or broccoli. Let squash vines run among a block of cornstalks. Shade a summer crop of lettuce or spinach by planting it on the north side of a trellis of beans. By observing how your favorite vegetables grow in your climate, you can invent your own custom designs for intensive planting.

Dividing the planting area into squares lets you plant in intensive blocks for maximum production.

1 Before planting, divide the planting area with string into a grid of squares so that you can position plants the same distance apart in all directions.

2 Dig planting holes, then remove a seedling from its container. Gently grasp the plant's leaves, and support it under the roots with a wooden plant label or an old teaspoon.

3 Set the plant in the hole so that it will be positioned at approximately the same depth it was growing in the container. Firm soil carefully around the roots.

4 This planting pattern enables you to fit more plants into the bed than traditional single rows allow. When the bed is planted, remove the string and water well.

Edible Landscapes

Decorative Vegetables

Some of the prettiest vegetables have bright-colored leaves, flowers, or fruits. Plants like ruby chard, pink or lilac ornamental kale, striped eggplants, or red, gold, orange, purple, or chocolate brown peppers make cheerful accents that break the monotony of plain green foliage.

An edible landscape is both attractive and productive. From a distance, it may resemble a conventional landscape, with colorful beds and borders surrounding a lawn or patio, perhaps enclosed by a hedge or accented by some specimen shrubs or trees. Looking closer, though, you'll see that the trees and shrubs bear fruit and berries, and the beds are filled with a mosaic of vegetables, herbs, and edible flowers. The element of surprise, of finding unexpected beauty in common food plants, is part of the fun of making an edible garden.

Designing an edible landscape is a creative solution to the problem of limited space. It allows you to enjoy having homegrown vegetables even if you don't have space for a designated vegetable garden. But there's more to making an edible landscape than just planting a patch of vegetables in the front yard.

First, think about the overall design, what you want, and what you have. Do you want a feeling of privacy and enclosure, screened from the world outside, or a friendly open garden that welcomes neighbors and visitors? Do you want to create a display that delights passersby, or a scene to enjoy from inside the house? Are there existing plants and features to work around, or will you be starting from scratch on a new lot?

Don't plan to fill the landscape solely with vegetables. Use fruit trees, berrying shrubs, and evergreen herbs to provide summer shade and screening and winter form and interest. Build fences and paths. Make an arbor or trellis. Choose a sheltered place for a seat or bench. Put up a sundial or a bird bath. These elements will establish the form and style of

Eggplant's glossy fruits and lavender flowers are decorative in the garden.

Most bell peppers eventually ripen to red if the season is long enough, but some varieties are purple or orange beforehand.

Red-leaved lettuce is available in looseleaf and heading varieties.

the landscape and make a permanent background for a changing seasonal display of vegetables.

Four of the best vegetables for edible landscapes are perennials that come back each year. Globe artichokes are giant thistlelike plants that make dramatic specimens. Jerusalem artichokes form a spreading patch of erect stalks topped with cheerful yellow sunflowers in fall. Asparagus sends up tender new shoots in early spring, then matures into a tall bushy fountain of delicate green foliage that turns gold in late fall. Rhubarb makes a broad mound of wide glossy leaves, topped with a large plume of tiny white flowers in early summer.

Cherry tomatoes, sweet or hot peppers, snap peas, okra, summer squash, cucumbers, scarlet runner beans, and climbing or pole snap beans are top candidates for edible landscapes, because they keep producing for weeks and weeks if you pick them regularly. Most of these vegetables have foliage that stays healthy and attractive all season, too. Chard, kale, New Zealand spinach, sprouting broccoli, and Brussels sprouts are also long performers. Sweet potatoes are vigorous vines that will cover a tepee, trellis, or fence with a thick screen of heart-shaped leaves from early summer to first frost, when you must promptly dig up the tubers.

In the cooler weather of spring and fall, plant a salad garden of greens, herbs, and edible flowers. Spinach, looseleaf lettuce, endive, and other mixed greens grow fast and recover quickly from repeated harvests. Use them to edge a bed of French sorrel, chives and garlic chives, parsley, dill, nasturtiums, calendulas, and pansies.

The bright red stems of chard stand out in an edible landscape.

Purple snap beans are especially tender and mild-flavored. The vines bear pretty flowers of lighter purple.

Scarlet runner beans do particularly well where summer is not terribly hot and dry.

Edible Landscapes CONTINUED

Bright flowers in assorted colors enliven this sumptuous edible landscape, which surrounds a central bed of salad greens, root crops, herbs, and broccoli.

In an urban backyard, ter-raced beds of flowers and vegetables turn a small space into a lush oasis.

Red-stemmed chard, blos-soming chives and borage, lettuce, spinach, and pansies combine in a pretty little pocket garden. All the plants are edible, including the pan-sies and borage blossoms.

This neatly designed garden features curved rows of let-tuce and beds of flowers in shades of green, pink, and purple.

Vegetables in Containers

Don't overlook containers when you think about places to grow vegetables. If you have very little (or no) space for a vegetable garden, containers offer an ideal solution. A sunny patio, city rooftop, or fire escape can accommodate as many contained vegetables as the space and weight-bearing capacity of the structure allow. If you have a lot of shade on your property, you can take advantage of a small sunny pocket in the corner of the yard, on the deck, or along the driveway and grow a few pots of tomatoes, lettuce, or eggplant. You can even use seasonal sunlight by shifting potted vegetables to keep them in the brightest light and avoid changing shade patterns as the sun moves farther south over the course of the summer.

Almost any kind of vegetable will grow happily in a pot as long as the pot is big enough to accommo-date its root system. Even vining cucumbers and zucchini will take to container life if you fasten the vines to a trellis and support heavy fruit with cloth slings. But the best bets for container growing are vegetables with a more confined habit of growth. Good possibilities include a range of salad greens, spinach, Swiss chard, beets, carrots (especially half-long and round-rooted varieties), bush beans, peppers, eggplants, determinate tomatoes, and bush varieties of squash and cucumbers.

All sorts of containers are suitable for vegetables, as long as they allow good drainage. Unglazed clay pots come in many sizes, and there are also terra-cotta boxes, bowls, strawberry jars, urns, and rectangular window-box-like planters. Clay containers allow air to pass through their porous walls, letting oxygen reach the roots. The drawbacks to clay are that pots

Containers of vegetables can turn an unused corner of a sunny patio into productive space. Group the pots to make watering easier and to create the look of a garden.

can dry out very quickly on hot, windy days, and large pots full of moist soil are quite heavy.

Many gardeners prefer plastic containers, which are lighter and don't dry out quite as fast as clay. They, too, come in a host of sizes and colors.

For large plants, or groups of plants, consider half-barrels, tubs, boxes, and other large planters made of wood, fiberglass, pressed fiber, or plastic resin. Lining wooden containers with heavy plastic before filling them with soil mix will make them last longer. Remember to punch drainage holes in the plastic at the bottom of the container.

To create the effect of a garden, group containers together and stagger the plant heights as you would in a bed or border. Place smaller pots of petite leafy greens, beets, or carrots in front of large tubs holding taller plants such as staked tomatoes, eggplants, or trellised peas, or set pots on staged shelving or tiered stands to create several height levels.

Or you could consider grouping several plants in a big half-barrel or box-type planter. Put a tall plant like an eggplant or pepper in the center (if the container is freestanding) or in the back (if the container will be viewed from one side). Smaller plants like leaf lettuce and arugula can be placed toward the edges of the pot.

Contained vegetables can also serve a variety of landscape purposes when you apply some creative thinking. For example, you could use a row of tall plants in large pots to create a screen or space divider. Consider a row of tall sunflowers or pretty purple-pod snap beans in tubs with trellises installed. Or try planting some scarlet runner beans in pots and training them around a porch.

Plant your potted vegetables in a light, porous soil mix such as the one found in the tip on this page.

Vegetables in containers will need water often in hot weather; water small pots thoroughly once a day or even more. If you are away from home often and you have a lot of container plants, you may want to invest in an automatic watering system. You will also have to fertilize frequently—apply a water-soluble fertilizer every week or two, depending on the nutrient needs of the vegetables you are growing.

EARTH·WISE TIP

A good all-purpose soil mix for vegetables in containers is equal parts of potting soil or garden soil, finely crumbled compost or leaf mold, and vermiculite, peat moss, or perlite.

A large tub or barrel can accommodate a group of plants. In this partially shaded garden, cabbage keeps company with white sweet William and purple violas.

Seasonal Crops

Vegetables
That Tolerate Frost

Plant seeds of peas, spinach, lettuce, beets, carrots, and radishes as soon as you can work the ground in spring. These vegetables prefer cool weather and tolerate cold soil and frosty nights. Some cole crops, such as cabbage and cauliflower, can take light frosts but may be stunted by temperatures of 27° to 28°F and below. Other cabbage family crops, such as kale, collards, and Chinese cabbage, can stand the lower temperatures without any problems.

Shopping in the produce aisle of a modern supermarket, you might get the idea that all kinds of vegetables grow year-round. The selection is about the same whether you're planning Christmas dinner or a Fourth of July picnic. Week after week, there are artichokes and asparagus, beans and broccoli, celery and cucumbers, peas and peppers—all lined up side by side, just as if they'd grown up together in one big happy garden.

But, of course, this is an illusion. Different vegetables don't all ripen at once. Some prefer cool weather. Others thrive in the heat. The only way a supermarket can offer a wide variety of fresh vegetables year-round is by shipping them in from distant places with different growing seasons.

Growing your own vegetables is a whole other matter. Instead of the timeless sameness of the supermarket, a home vegetable garden offers an annual

This garden is already full of vegetables when the tulips bloom. Sugar snap peas climb a trellis, while the central bed contains red mustard and assorted other leafy crops.

Colorful tomatoes and peppers are classic summer vegetables.
Just a few plants can yield a bounty.

cycle of seasonal fare. No garden can match a super-market for variety on any one day, but anticipating and savoring each crop in turn gives variety to the year as a whole. The first green salad grown in your own garden is a wonderful tonic in spring. In summer, biting into the first ripe tomato or ear of corn is such a special event that you want to mark the date on the calendar. Yet when those crops are past, you turn gladly to the fall flavors of buttercup squash, leeks, and Brussels sprouts.

Watching the weather map on the evening news is a daily reminder of how much the weather varies from region to region around the United States. It takes years of experience to learn just when to plant each crop for the best results in your particular garden, but in general, we can divide the country into

regions of cool or warm climates, and divide vegetables into cool-season or warm-season crops.

Cool-climate regions include New England, the mid-Atlantic, the Great Lakes, the Midwest, and the western mountains. In these regions, spring starts when the ground thaws and continues until the soil is warm. Varying from place to place and year to year, spring can be as short as two to three weeks or as long as two to three months. Don't wait for summer to begin your garden. Get started as early as possible with cool-season crops such as peas, fava beans, spinach, lettuce, carrots, beets, radishes, onions, potatoes, cauliflower, broccoli, and cabbage. Spring is also the time to plant slow-growing crops that need several months to develop, such as parsnips, salsify, scorzonera, celery, celeriac, and leeks.

Seasonal Crops CONTINUED

TROUBLESHOOTING TIP

Vegetables ripen fast during the heat of midsummer. If you're going to be away for a week's vacation, invite friends and neighbors to stop by and pick what ripens while you're gone.

Spring peas, carrots, and salad greens reach their peak harvest in early summer, while corn and other summer vegetables continue to grow.

Summer is devoted to warm-season vegetables such as tomatoes, beans, corn, and squash. Everyone grows them, and you should, too. But save a little space for mid- or late-summer plantings of cool-season crops that will carry over into the fall—the season after the first frost and before the ground freezes. (Like spring, fall can be brief or prolonged.) Spinach, lettuce, and other salad greens, as well as Brussels sprouts, cabbage, kale, and other cole crops, all develop fine fla-vor, appearance, and quality in the short days and cool nights of fall. Carrots, beets, and turnips can be left in the garden until the ground freezes; if protected with mulch, they keep all winter.

Warm-climate regions in this country include the West Coast, Southwest, and southern states from California to Texas to Georgia and parts of the Carolinas. There the ground rarely freezes in winter. Fall plantings of salad greens, cole crops, and root

Some vegetables develop their best flavor and quality during the short days and cool nights of fall. Lettuce, spinach, and other greens are very crisp and succulent. Leeks grow thick and tender, and carrots become incredibly sweet and juicy. Kale, cauliflower, Brussels sprouts, and other cole crops also taste sweeter after they've been touched by frost.

Plant a second crop—or several successions—of beets and radishes to harvest in fall.

crops can carry over into the new year, and spring planting can begin in January or February. In warm climates, the challenge is taking advantage of the long frost-free season. Only a few kinds of vegetable plants, like chili peppers and okra, will continue to grow and bear produce through months of hot weather. Most other warm-season crops, such as corn, beans, tomatoes, and squash, get used up or worn out, and must be replanted at monthly intervals to supply continuous harvest. Many warm-climate gardeners take a break during the hottest part of summer, and start planting with renewed enthusiasm when the weather cools in fall.

Succession Planting and Crop Rotation

Succession planting is a strategy for prolonging the harvest. The tactics vary with different crops, but the goal is the same: to produce more vegetables over a longer season.

In some cases, the potential growing season for a crop is longer than the life span of the individual plants. For example, radishes are most tender and tastiest about four to six weeks after planting; after that, they get tough and peppery. But the season of cool weather that radishes prefer may extend for up to three months. The best way to enjoy fresh radishes throughout this season is to plant small crops at regular intervals, every week or two, and harvest each crop when it reaches its prime. This tactic applies to many other fast-growing crops, such as kohlrabi, beets, leaf lettuce, turnips, and spinach.

Another tactic is convenient when a crop has different varieties that mature at different rates. You can plant several varieties at once, then harvest them in sequence. For example, early cabbages are ready to pick about 60 to 65 days after you set out the plants, but late cabbages require 70 to 90 days. Early sweet corn ripens about 65 to 70 days after sowing, but later varieties take 80 to 90 days. You might wonder if it's worthwhile sowing the slower-growing varieties. Why not make successive plantings of the faster-growing kinds, instead? Sometimes that's a good idea. But in many cases, the late varieties such as Silver Queen corn, Big Boy tomatoes, or Savoy cabbage have better flavor, size, and quality than earlier kinds.

Related to succession planting is the idea of keeping the garden full at all times. Empty space is wasted opportunity. Holes are formed wherever crops yield just one picking, like carrots or cauliflower; or mature

1 *Succession planting makes the garden more productive. When an early crop stops producing, pull up the plants and put them on the compost pile.*

over a short season, like peas or sweet corn. Some gardeners start seedlings in flats or a nursery bed and have them ready for filling small gaps as they open. Others pull one crop, till and rake the area, then plant a different crop. You might design an arrangement where melons or squash could sprawl to fill the gaps left by earlier crops.

Even if you don't have the time or desire to plant more vegetables, don't let empty spaces be taken over by weeds. It's much better to fill them with cover crops or green manures—plants that grow fast and produce lots of organic matter for composting and soil improvement, but are easily managed and don't spread like weeds. White clover, sweet clover, alfalfa, hairy vetch, buckwheat, oats, and winter rye are among the many options. Any can be seeded directly to make good use of small or large gaps.

2 *Add fertilizer or compost to the bed if you wish. Then rake the soil, removing any large rocks, to create a level, fine-textured planting area.*

3 *Plant transplants or sow seeds of the new crop. Use stakes and string to make row guides to keep plantings in a straight line.*

EARTH•WISE
TIP

Plant legumes, such as beans and peas, to enrich your soil. The roots of these vegetables are host to soil bacteria capable of extracting nitrogen from the air. Thus, in addition to providing a harvest for your table, the legumes also store nitrogen in the soil for other crops.

Crop rotation is like a game of musical chairs. It involves moving crops around the garden, trying not to grow the same or similar plants in the same place time after time. One purpose is to avoid troublesome pests and diseases that harbor in the soil. Another reason is that crops use nutrients in different ways, so crop rotation is important for maintaining balanced soil fertility throughout the garden. If you can spare the space, including cover crops in a rotation scheme is an excellent way to improve the soil, and it also helps combat disease and control weeds.

The easiest way to plan crop rotations is to group vegetables by plant families. In general, members of a family are susceptible to the same pests and diseases and use similar nutrients. From season to season and year to year, whenever you plant a new crop, try to choose a vegetable from a different family from what was growing there previously. For example, follow members of the nightshade family (eggplants, peppers, potatoes, and tomatoes) with something from the mustard family (cabbage and other cole crops), or the legume family (all kinds of peas and beans). Two or three times a year, sketch a simple map or planting diagram of your garden and indicate what's planted where. Then you can refer to older maps when planning future layouts.

Mediterranean Garden

W̶ho can resist the aromas and flavors that characterize the gardens of the Mediterranean? The traditional vegetables and herbs of the region are ideal companions in the garden as well as in the pot.

This small-space garden packs in all of the fundamental tastes of the Mediterranean. Tomatoes, peppers, onions, garlic, and other vegetables grow intermixed with the traditional herb flavors of basil and oregano. Ideally, the garden should be located only a short distance from the kitchen, so that it is easy to capture the very freshest tastes.

This garden's architecture provides a framework for planting for years and years, although its permanent nature requires a lot of careful planning. The steps are of cut stone, and the tiles can be plain or painted. Even when the garden is put to bed for the winter, this will still be an appealing spot.

The stepped walkway, which is evocative of Italian garden architecture, provides an effective garden design for a sloping terrain. Terraced gardens are found throughout the Mediterranean region, testimony to the determination of the people to garden despite the topographic conditions. Tomatoes, as well as other plants featured here, grow well on terraces because of the benefits of good drainage and warm soils, particularly on south-facing slopes.

Plant List

1 Salad tomatoes
2 Paste (and pear) tomatoes
3 Zucchini
4 Eggplants
5 Broccoli rabe
6 Fennel
7 Assorted peppers
8 Onions
9 Garlic
10 Parsley
11 Romaine
12 Arugula
13 Radicchio
14 Basil (in pot)
15 Oregano (in pot)

Putting basil and oregano in pots allows you to bring these tender herbs in on cool nights, avoiding the possibility of their being killed by an early frost. Move the pots out again the next day, after the temperature rises.

Colonial Garden

*H*ere's a nostalgic garden that may remind you of colonial days. But it's still a useful and attractive garden for today's serious vegetable gardener.

From the nodding sunflowers to the pumpkin vines and cornstalks, this garden speaks of the joys of the harvest. Many of these vegetables, particularly the root crops, will keep well into the winter if stored properly in a cool, dry spot.

Feel free to indulge yourself in perennial vegetables with this garden. Sorrel, horseradish, and rhubarb were treasured in colonial Americans' gardens. Once established in a family garden, they were kept going for years and years.

Plant List

1 Sunflowers
2 Corn
3 Rhubarb
4 Sorrel
5 Onions
6 Leeks
7 Parsnips
8 Horseradish
9 Cabbage
10 Potatoes
11 Turnips
12 Beets
13 Winter squash
14 Pumpkins
15 Pole beans
16 Peas
17 Brussels sprouts
18 Carrots
19 Tomato tepee
20 Red cabbage
21 Lettuce

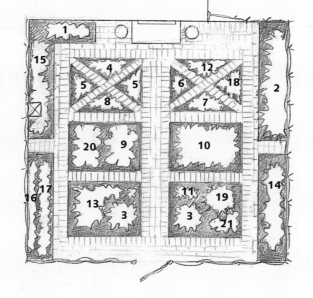

The old-fashioned wattle fence surrounding this garden is actually fairly easy to make. It is formed from brush and tree saplings that

Vary your selection of plants to meet your tastes, trying new things from year to year. It's a good idea, too, to rotate the location of your plants annually. Different vegetables deplete the soil of different nutrients. Crop rotation restores balance to the soil. The periodic addition of compost will also help to enrich the soil. Don't forget to keep your paths wide enough for your wheelbarrow—you'll have big harvests in this garden!

are horizontally woven between larger, upright saplings, utilizing the same principles as basket weaving. It's great for keeping out pests, too, unlike fences with more space in between the rails. In the old days, gardeners used long poles for the uprights and when they rotted, they'd just bang them farther into the ground.

Pick and Eat Crudités Garden

*J*ust a step away from the hors d'oeuvre tray, a "pick and eat" garden satisfies the eye as well as the palate. This well-tended vegetable garden is as pretty to look at as a more traditional flower garden—and it's full of delicious salad ingredients.

Here is the perfect way to have an edible garden without committing to the traditional rectangular plot in the middle of the yard. Designed more like an ornamental border than a vegetable garden, this land-scape gracefully separates the lawn from the patio.

Don't forget to add some nasturtium blossoms to the top of your salad—they're colorful and have a delicate, peppery taste.

Kale is decorative and highly nutritious. A little bit goes a long way—to garnish a platter or contribute to a mixed green salad. You can pinch off outer leaves and kale will keep on growing from the center—and it will thrive past early frosts right through into the winter.

The patio flagstones extend into the garden, providing easy access to the many varieties of vegetables and herbs. Weeding will be easier if you don't have to lean over more than 3 feet from the walk-way or grass to reach the center of the garden bed. To cut down on weeds, mulch between plants; an attractive landscape mulch will look good from the patio and will also help to retain needed moisture in the soil.

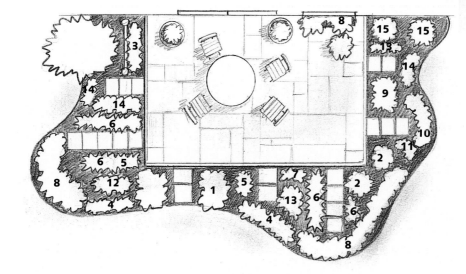

Plant List

1 Flowering kale
2 Cherry tomatoes
3 Snap beans
4 Spinach
5 Corn-salad
6 Onions
7 Radishes
8 Nasturtiums
9 Baby carrots
10 Arugula
11 Cress
12 Mustard
13 Romaine
14 Parsley
15 Tomatoes

If you do a lot of entertaining on the back patio, consider adding some subtle garden lighting, hidden among the plants, to extend the beauty of this garden into the evening hours. Even if you've retired indoors after dark, the garden will be lovely to look at if softly lit. If you do light the garden, substitute a different leafy vegetable for the spinach, which grows best when days are short. The additional light may cause spinach to bolt to seed.

Neat and Easy Raised Beds

It's also possible to create waist-high garden beds, which are particularly welcome to gardeners who have back trouble or who are confined to a wheelchair. Be sure to keep the distance between beds wide enough to allow a wheelchair easy access and room to turn around.

*R*aised beds provide an easy and effective method for creating a backyard vegetable garden. Instead of tilling the soil, you build a garden bed above ground level. "Raised beds" is simply a term used when the soil is raised up so that the seedbed is higher than the pathway or the surrounding ground level.

If you have wet or heavy clay soil, raised beds are a particularly good idea because the soil drains better, and good drainage helps to prevent plant disease. Also, in raised-bed gardens, the soil warms up earlier in the season—a definite advantage for northern gardeners with a short growing season. Orient the garden with the trellis on the north side, where it won't shade the other plants—unless you live in a hot climate, where you might want the shade.

To get the most effective use of space in a raised-bed garden, create wide rows as shown here. The depth of the soil should be at least 8 inches, with the top few inches well cultivated for planting seeds. If you plan to grow root crops, increase the depth to at least 12 inches. Keep the two beds the width of a mower apart for easy and efficient mowing. Note the mowing strip surrounding each bed. This strip is a ground-level board that is wide enough for the wheel of the mower to go over, eliminating the need to hand-trim around the beds.

Plant List
1 Cucumbers (on trellis)
2 Scarlet runner beans (on trellis)
3 Cherry tomatoes (on trellis)
4 Bibb lettuce
5 Ruby Swiss chard
6 Spinach
7 Bush zucchini
8 Broccoli
9 Carrots
10 Assorted peppers
11 Onion
12 Red leaf lettuce
13 Leaf lettuce

The trellises have widely spaced openings, so that you can pick from either side, and are not so high that you can't reach the top. These trellises are braced to the beds' rear boards, so that wind or the weight of plants won't topple them. Ask a lumberyard for suggestions about the best wood to use, but avoid using pressure-treated lumber that has been processed with caustic chemicals, since your vegetables will be in contact with the wood.

Managing the Garden Environment

*l*earning to manage the garden environment, even if some things are beyond human control, is a key aspect of gardening. Obviously, you can't make the sun come out on a cloudy day or stop a spell of rainy weather. But you can do a lot to improve the growing conditions for your plants. • The first step in garden management is providing good soil to support and nourish your plants. Even if your site is now barren, you can soon create a deep bed of healthy soil. Choose a location that gets plenty of light, but take steps to protect sensitive plants from too much sun. Also, be sure to compensate for irregular rainfall by improving drainage so that excess water soaks away or by irrigating when the soil gets too dry. • Managing a garden takes time and skill, but any gardener will agree that the rewards far exceed the effort you put into it.

Soil

TIMESAVING TIP

Thick stems such as cornstalks or squash vines compost faster if you chop them into short pieces. If you don't have a chipper or shredder tool, use hedge loppers to slice them up, or spread them on the ground and run back and forth with a rotary mower.

More than any other group of gardeners, vegetable growers are passionate about soil. That's because vegetables respond so quickly and dramatically to your efforts at soil improvement. Every step you take shows in the increased health, vigor, and productivity of your crops.

Few gardeners are lucky enough to start off with good soil. More often, the soil on a potential garden site is shallow, compacted, infertile, and lifeless. It might support a straggly lawn or some tough weeds, but not a vegetable garden. For vegetables, the ideal soil is deep, loose, rich, spongy with organic matter, and teeming with earthworms and microorganisms.

Fortunately, you can create wonderful soil, starting from scratch, on almost any site, and the process only takes a few years. All it requires is vision, optimism, some hard work, and lots of organic matter. Gardeners from coast to coast have built beautiful, bountiful vegetable gardens on even the most sterile and challenging sites—former parking lots, abandoned inner city blocks, suburban subdivisions where developers scalped the topsoil, and hillsides eroded by decades of careless farming. With luck, your situation will be much easier.

Whether you're starting on a vacant lot, digging up an area of lawn, or developing an existing garden, the first step in soil improvement is staking out the site and digging or tilling the entire area to break through the surface and expose buried objects. Lift any healthy sod and transplant it to other parts of the property, or pile it upside down to compost. Remove weeds, roots and all, and spread them in the sun until they are thoroughly desiccated (so they won't sprout new growth) and add them to the compost pile. Pull out all the rocks, chunks of concrete, old construction debris, cans and bottles, and other junk you find

1 *To make compost, build a pile with alternating layers of wet materials such as green plant debris and dry materials such as shredded dry leaves. Add a thin layer of soil.*

buried in the soil. Make at least one more pass over the entire area, digging as deep in the ground as your equipment will go. The goal is to break through dense layers and loosen the soil.

Doing this much work is likely to attract the attention of your neighbors, especially if you're a newcomer. Use the opportunity to find out if anyone else grows vegetables, and ask their opinion of the local soil. Also, this is the time to collect a soil sample for testing. Most states offer inexpensive soil tests through the county cooperative extension service. When you send in the sample, tell them you want to make a vegetable garden. The test results will tell whether your soil is acid or alkaline, and may recommend adding lime to raise the pH level or sulfur to lower it. It takes a while for lime to have an effect, so apply it as soon as possible and work it deep into the

2 *Kitchen wastes such as vegetable peels, spoiled fruit, and coffee grounds are excellent "wet" compost ingredients. Store scraps until you have enough to make a layer.*

3 *Turn the pile over with a pitchfork every week or two to hasten decomposition. Compost is ready to use when it is uniformly dark brown and crumbly, with an earthy smell.*

soil. Sulfur works faster, so you can wait until planting time and apply it to the surface. (The test will also report the levels of nitrogen, phosphorus, and potassium in the soil, and will recommend how much fertilizer to add. For more about fertilizing, see pages 80–81.)

Now the hard work is done and the fun part begins. In almost all cases, new garden soil is deficient in organic matter. It needs a stew of decomposed plant and animal residues to help improve the soil texture, soak up moisture, and supply plants with small but steady doses of vital nutrients. Increasing the amount of organic matter in your garden's soil quickly improves how well plants grow.

You can buy organic matter, scrounge around for free (or at least cheap) ingredients, or produce your own. Any garden center sells bales or bags of peat

This compost pile was constructed from layers of loose soil, straw, grass clippings, tree litter, fresh weeds and prunings, old leaves, and alfalfa pellets.

moss and composted steer manure. Many towns now sell or give away compost made of leaves, chipped brush, and grass clippings. Farmers sell or give away stable manure and bedding and spoiled hay or straw. Sawmills make piles of bark and sawdust. You can make your own compost from yard and garden wastes and kitchen scraps, and you can plant special crops, called cover crops or green manures, specifically to produce organic matter.

In every case, the same general guidelines apply. If the organic matter has decomposed into dark, amorphous, unidentifiable crumbs, it's good for mixing into your soil. If you can still recognize the original material—leaves, stems, or other plant parts—don't bury it yet. Compost it, or use it for mulch, but don't mix it down into the soil. Freshly gathered organic matter, especially garden debris, can harbor various pests and diseases. These problems are reduced or eliminated as the material breaks down, but decom-

position takes place slowly underground because the microbes that do the work need plenty of oxygen. It is best to let the material decompose aboveground, where it happens much faster.

There are many ways to make compost. The easiest but slowest way is simply to pile up the material you have and let it sit for a year or so. It helps to turn a pile once or twice during the year, tossing the rougher material from the top and sides of the old pile into a new mound, then covering it with the more-decomposed material from the center and bottom. When it looks like dark chunks of soil, the compost is ready to use. This method is called cold composting, because the pile never heats up. It's fine for composting sod, hedge trimmings, autumn leaves, wood chips, sawdust, and similar raw materials that don't contain weed seeds or disease organisms.

Hot composting is better than cold composting for lawn clippings, hay, stable manure, kitchen scraps, and garden waste such as bean vines or broccoli stems. In hot composting, vigorous microbe activity rapidly heats the contents of the pile to about 160°F, hot enough to kill most weed seeds and disease organisms. Another advantage is that decomposition happens faster at high temperatures, giving you finished compost in a matter of weeks, not months. One problem, though, is that you need a lot of material to start with—1 cubic yard is the minimum, and 2 cubic yards are better. Smaller piles just don't heat up enough. One solution is to stockpile ingredients before adding them to the compost. Let green plant parts dry in the sun before you collect them, then store the dry material under a plastic tarp until you have enough to make a hot pile. Bear in mind, to keep the pile hot you need to turn it often—at least once a week—so decomposition occurs quickly.

How Earthworms Help the Soil

Earthworms are a gardener's best friend. They tunnel up and down through the soil, opening passageways for air and water to flow freely. They carry organic matter underground and mix it with the soil, releasing nutrients for plants to absorb and churning out granular pellets that improve the soil's texture. The presence of earthworms is an indicator of good soil.

When you're ready, prepare the pile by layering different ingredients. Stick a few smooth poles or pipes like birthday candles in the bottom layer, stack material around them, and when you're done, pull out the poles to leave vertical air columns in the pile. If possible, include one part of fresh cow or horse manure for each two or three parts of plant material, by volume. Adding manure makes the pile heat up faster and better. Moisten the material as you layer the pile, but don't get it soggy wet. Don't pack it down, either—a loose pile includes more essential oxygen.

The pile should start heating within a day or two. The best way to monitor its temperature is with a long compost thermometer, available at many garden centers, that you can push into the center of the pile. When the temperature gets to about 150°F, turn the pile, inverting its contents and moistening any materials that feel dry. (Remember to make air holes with the poles.) Then wait for the pile to heat up again. After two or three turnings, the material should be thoroughly mixed and the decomposition well underway. When the pile cools off to below 100°F, the compost is ready to use. If you don't use the compost right away, keep it covered so rain doesn't leach away desirable nutrients.

You can make compost in a pile on the ground, but for neatness, it helps to build a bin with scrap lumber. Even a simple ring of snow fence or woven wire fencing will help shape a more tidy pile. Several models of stationary or rotating compost bins are available from garden-supply catalogues. These are ideal for disposing of kitchen scraps and small amounts of yard and garden waste, but they don't hold enough to supply much compost for a large garden.

To create a fine-textured bed for seed sowing, press soil through a screen or soil sieve to remove stones and break up clumps.

How much organic matter should you add to the soil in your vegetable garden? Some gardeners would say you can never add too much, and it's true that plants thrive in pure compost. But in practical terms, how much you apply is usually limited by what's available or what you can afford to buy. If you're starting a new garden in typically lean soil, try to spread a 3- or 4-inch layer of well-composted organic matter across the soil and work it in to a depth of 8 to 12 inches. In better soil or in subsequent years, apply a thinner layer, just 1 or 2 inches deep.

Another way to add organic matter to your soil is to plant fast-growing cover crops, sometimes called green manures. This is a good way to start a new garden or renew an older one. Cover crops are sown at close spacing and quickly produce a lush cover of top growth and an extensive root system.

Soil CONTINUED

Some cover crops grow best in cool weather and are usually planted in the fall. This group includes winter rye, wheat, barley, hairy vetch, and crimson clover. Others do best in warm weather and are usually planted in the spring. These include oats, red clover, sweet clover, alfalfa, and buckwheat. White clover can be planted in any season.

To plant a cover crop, dig or till the soil, then rake the surface smooth. Use a lawn grass seeder, or walk back and forth to sow the seeds by hand, spreading them as evenly as possible. Rake lightly to cover the seeds with a thin layer of soil. Water frequently until all the seeds have sprouted and the seedlings are a few inches tall.

Within a few months after planting, you can mow off the tops, then dig or till to mix all the clippings, stubble, and roots into the soil. Wait at least a week in warm weather, or longer in cool weather, for the residue to begin to decompose before you plant another crop.

As this chapter has repeatedly suggested, digging or tilling is part of establishing good soil for vegetables. To create a new garden, a strong back and a stout digging fork are usually required, and a gas-powered rotary tiller is a welcome ally. After the first year or two, though, the work gets easier. Maintaining healthy soil is much easier than creating it.

In fact, many vegetable gardeners practice a system called no-till gardening. After the first year of soil preparation, they make a point of not tilling or digging, because they want to preserve, not disturb, the horizontal layering of the soil. They copy nature's ways, adding organic mulch to the surface and letting earthworms and other tiny critters chew it to bits and carry it down and mix it into the soil. No-till gardeners are careful never to step off the paths and compact the soil in their planting beds, and they practice crop rotation, so that different plants will root into different layers of the soil from year to year. It's a care-free system that can work very well.

Zucchini and other summer squash grow best in soil that is well drained, slightly acid, and rich in organic matter.

Healthy soil is the basis of a good garden. A cover crop of red clover (foreground) will be tilled into the soil to add valuable organic matter.

Light

1 *You can postpone bolting and extend the lettuce harvest a couple of weeks by keeping the soil moist and shading plants during the hottest part of the day.*

2 *Cover the plants with shade netting, floating row covers, or old screen doors or windows from late morning to mid-afternoon, when the sun is hottest.*

All plants collect energy from light and, through photosynthesis, use it to grow. Vegetables in particular need plenty of light in order to grow fast, stay healthy, and yield abundant crops.

It's often said that vegetables need full sun, but that means different things in different climates. Regional guidelines are more specific. Across most of the country, where summers aren't too hot and there are more clear than cloudy days, most vegetables need at least eight hours a day of direct, unshaded sun. However, in Florida, Texas, the Southwest, and other areas where the summer sun is very hot and bright, six hours of direct sun is enough for crops. And in coastal areas like San Francisco, where the weather is often foggy, vegetables need all the sun they can get, whenever they can get it.

In general, too much sun, especially hot afternoon sun, makes vegetable plants lose water faster than the roots can absorb it. This leads to wilting, but if the soil is dampened, the wilted leaves will recover overnight and look fresh and crisp again by dawn. If the soil remains dry for too long, the effect is more serious: the wilted leaves may look blistered or spotty, turn gray or tan around the edge, and shrivel. By the time these symptoms appear, the plant is getting weak. It may not die right away, but its growth and yield will be reduced.

Leafy crops are especially sensitive to too much sun. Southern and desert gardeners should use a canopy of lightweight spunbonded fabric or a lean-to lath shelter to protect a patch of lettuce, spinach, and other greens for summer salads. New transplants of any crop are also vulnerable and need shading from intense sun for the first week or so. Some gardeners stick a shingle or short board in an upright position on the south side of a transplant. Others use long, leafy hedge trimmings to make a temporary tepee. You can even unfold the bottom of a paper grocery sack, turn out the flaps, and use rocks to hold it in place around a plant.

Sunburn or scald is a common injury on green peppers and tomatoes, where it appears as large translucent spots that are weak, thin, and subject to further injury or infection. Normally these crops are protected from sun by the plants' foliage, but sunburn is likely if foliage is removed by disease, storm damage, or pruning. In these cases, use lightweight spunbonded fabric to shade the developing fruits until the foliage fills in again.

Too little sun is even more detrimental than too much sun. Few vegetables grow well in the shade of trees, other tall plants, or buildings. If your property is shaded but you really want to grow vegetables, consider removing a few trees or shrubs to let in more light. You might try positioning a large mirror or a piece of plywood painted glossy white to reflect light back into a shady corner. Gardeners in cloudy climates have reported good results from using aluminum foil as a mulch between rows to reflect light back up to the bottom of the leaves.

When we talk about too much or too little sun, we're referring to light intensity. A related topic is day length—the hours of sunlight from dawn to dusk. Day length has specific effects on a few kinds of vegetables. For example, some onions form bulbs only when the days are long, but Jerusalem artichokes don't make tubers until the days start getting shorter in fall.

Shade-Tolerant Vegetables

If your garden gets less than six hours of sun a day, plant leafy vegetables such as lettuce (top left and right), spinach (bottom right), Swiss chard (top center), endive, cabbage (bottom left), escarole, cress, sorrel, and arugula. Carrots and beets will grow bigger tops than normal, but their roots will be smaller than normal. If there's space for the vines to stretch and sprawl, try pumpkins or winter squash.

Most vegetables respond to changing day length simply by growing faster or slower. Long days promote rapid growth. In northern Europe or Alaska, where the sun hardly sets in summer, vegetables grow incredibly fast and reach giant size with excellent flavor and tenderness.

Short days delay ripening. Even in a greenhouse, it takes weeks longer for a tomato to turn from green to red in winter than in summer. On the other hand, short days (and cool temperatures) provide ideal "holding" conditions for leafy vegetables such as salad greens, cole crops such as broccoli and cabbage, and root crops such as carrots and parsnips.

Water

Critical Watering Times

Dry soil is especially stressful for tiny plants whose roots are still close to the surface. Be especially careful to water regularly (once a day if needed) after sowing seeds, when seedlings are still small, or after setting out transplants in the garden. Fruiting crops also need water when they are flowering and setting their crop. Root vegetables need water as their roots are enlarging.

Vegetables need lots of water. Water constitutes as much as 90 percent of the fresh weight of many crops, and that's just in the part that is harvested, such as a head of broccoli or an ear of corn. In order to produce that part, the rest of the plant needs water, too. For example, one corn plant can use 60 gallons of water during the weeks from sowing to harvest. Providing enough water is one of the most important concerns in growing vegetables.

In most parts of the United States, vegetable plants require rainfall or irrigation averaging 1 inch per week throughout the summer growing season. In especially hot or windy weather or in desert regions, vegetables may need up to 2 inches of water a week. Crops growing in the cooler weather of spring, fall, or winter can get by with as little as 1/2 inch of water per week.

Count your blessings when rain comes in just the right amounts at just the right times. More often, there's too much or too little rain, too soon or too late. Gardeners, like farmers, have to live with the weather, but there are several steps that you can take to mitigate its impact.

First, consider the problem of too much rain. Excess moisture promotes fungus infections that rot plant roots and weaken and disfigure the stems and leaves. Too much water also causes mushy tomatoes that make insipid juice and runny sauce, pulpy potatoes that bruise easily and don't store well, beans and peas that sprout in the pod, carrots that crack open, and other disappointments. There's nothing you can do to stop the rain, but you can plan ahead and choose a well-drained site for your garden. Avoid places where water collects after storms or when the snow melts in spring. If your property has poor

1 *If you plan to install a drip-irrigation system, design the garden in straight lines. These raised beds with sloped sides will work.*

5 *Cut holes in the mulch for planting. Make the holes large enough for rainwater to penetrate between the mature plant stem and the plastic.*

2 Lay out drip tubing on top of the beds, and connect it with appropriate couplings. Run a drip line down the center of each narrow bed.

3 To grow warmth-loving crops, cover the tubing with soil, black plastic mulch, organic mulch, or even leave the drip lines exposed.

4 After stretching the plastic over each bed, weight down the edges with soil, or use metal pins or tent pegs to hold the mulch in place.

6 Plant a seedling (these are cucumbers) in each hole immediately after removing it from the container, so that the roots don't dry out.

7 Firm the soil around the roots by pressing on the top of the plastic—but do not pack the soil hard. Turn on the system to water after planting.

8 Finally, spread a thin layer of straw, wood chips, or other organic mulch over the area to hide the black plastic.

Water CONTINUED

drainage overall, build raised beds for growing vegetables. Cultivate deeply so that excess water will drain quickly into the soil, not linger near the surface. If you live in a rainy or humid climate, be sure to allow plenty of space between plants, so that air can circulate between and dry the foliage after a shower. And be careful about using hay, straw, grass clippings, or other organic mulches. In wet weather, these mulches retain so much moisture that they provide a haven for disease organisms and slimy pests such as slugs, snails, and pill bugs.

Too little rain is a more common problem than too much rain, but fortunately, it's easier to deal with. The first step is good soil preparation. Dig deeply and loosen the soil to a depth of 1 foot or more, so that roots can penetrate easily and tap water from a larger volume of soil. (Be careful not to walk on the soil and recompact it after you've loosened it.) Year after year, add new supplies of organic material to the soil. This acts like a sponge, absorbing rainfall or irrigation water and releasing it slowly to the plant roots. Use more organic material as mulch, to shade the soil from drying sun and wind and to conserve moisture in the root zone.

There are several ways to irrigate a garden. You can carry water in a bucket or watering can, hand-water with a hose, hook up a portable sprinkler, lay out soaker hoses, or install a drip-irrigation system. Each method has its pros and cons. Sprinkling with a can is quick and easy when all you need is to moisten a row of seedlings or settle a few transplants. Holding a hose works fine for watering containers or small gardens. Both are inexpensive but labor intensive.

Portable hose-end sprinklers are inexpensive, convenient, and available at any garden center or discount store. It takes just a minute to hook up a sprinkler and start watering. Set a few empty jars or cans in its path, and let it run until an inch of water has accumulated in the jars. The major disadvantage of these sprinklers is that some of the water is wasted to evaporation, and more is wasted where it falls on paths or unplanted areas.

Where water is scarce or expensive, it's worth investing in soaker hoses or a drip system. Both apply water directly to the root zone, where it's needed, with minimal waste or runoff. Some of the best soaker hoses are made from recycled rubber tires. They have a porous structure that allows a slow seepage of water, but doesn't clog with sediment as readily as hoses with just one row of tiny holes. You can lay a network of soaker hoses under the mulch and leave them in place all season, or just buy one and keep moving it around.

Drip-irrigation systems are the most efficient in terms of water use, but they require more planning—you have to decide where to run the main lines and connect the secondary lines—and there are lots of separate pieces to put together when installing a drip system. Follow the photos on pages 56–57 for step-by-step information on how to install a drip-irrigation system.

If you live in the Southwest, or elsewhere where summer weather is dry, consider planting in sunken—rather than raised—beds. Sunken beds capture and hold water for plants, and minimize water runoff.

Drought-Tolerant Vegetables

Once past the seedling stage, some vegetables can tolerate a spell of hot, dry weather because they have deep roots. These include asparagus, lima beans, tepary beans, pumpkins and winter squash, okra, Southern peas, chili peppers, peanuts, sweet potatoes, tomatoes, and watermelons. All do best if they are planted in soil that has been loosened to a depth of 1 foot or more and amended with plenty of organic matter.

CHAPTER 3

Growing Vegetables

growing vegetables is an interesting process because it involves so many different activities. • Every week in the season brings its own concerns. For many gardeners, the first step is sowing seeds indoors and nurturing tiny seedlings as they develop into sturdy transplants. Then comes preparing the garden for planting—tilling or forking, raking, staking out rows and beds, and deciding what to plant where. Planting goes quickly as you seed directly and transplant into the garden. • Then comes the season of watering, weeding, cultivating, mulching, watching for pests, and worrying like a farmer as you wait for the first vegetables to ripen. Picking is a job in itself, when a few packets of seeds have multiplied into bushels of harvest. • Finally the season will draw to a close, and you'll put the garden to bed and start planning for next year.

Starting from Seed

Most vegetables are grown from seed. In some cases, you can sow the seeds directly where you want the plants to grow. This is the best method for seedlings that grow especially quickly, or for plants that are difficult to transplant because they have a long, often brittle, taproot.

In other cases, it's better to sow the seeds in one place and let them grow for several weeks before transplanting the seedlings into their final location. This method is recommended for plants that develop more slowly and need extra care and protection while they are small. You can start seeds indoors under fluorescent lights, in a greenhouse, or in a cold frame or nursery bed in the garden.

The system of sowing and transplanting has several advantages. Sowing indoors or in a protected area means you can keep a close eye on the seeds and seedlings and protect them from cold, heat, drought, storms, wandering dogs, hungry birds, and other catastrophes. This is especially worthwhile for expensive new hybrids or rare heirloom seeds.

You can get a head start on the season by sowing seeds indoors weeks before the last frost; then, when the soil is warm enough for planting, you can begin with plants instead of seeds. This approach makes the most of a short growing season, and accelerates the harvest. Most gardeners do this for tomatoes and peppers, but it also works for eggplants, cucumbers, melons, squash, and many other crops. (Handle squash and melon seedlings carefully during transplanting to avoid damaging the roots.)

When the garden is full in summer, you can save space by starting seeds in flats or in a small nursery area, where they can grow for weeks before they need

When seedlings become crowded, thin them to allow continued growth until it's time to move plants into the garden. Some of the flats shown here have been thinned, while most of them are ready for thinning.

Recycled containers in which to start seedlings include eggshells, cell packs, and other pots in which garden center plants were purchased in a previous year.

When to Start Vegetables Indoors

Find the average date of the last spring frost in your area, then count back 10–12 weeks for onions and leeks; 8–10 weeks for peppers; 6–8 weeks for tomatoes, cole crops, and head lettuce; and 3–4 weeks for okra, squash, cucumbers, and melons.

to be transplanted to their final spacing. This is a good way to begin fall crops of cabbage, broccoli, and other cole crops.

Starting seeds indoors isn't difficult, and it's a lot of fun to watch the tiny plants grow. You can grow seedlings in any spare corner of the house or basement. For light, use an inexpensive shop-light fixture with two ordinary fluorescent bulbs, and an automatic timer to turn the light on for about 16 hours a day. (Lights are more reliable than windowsills, where cloudy days and long nights can retard growth.) Hang the fixture on adjustable chains, support it on blocks or bricks, or improvise another system for positioning it just a few inches above the soil to get started.

You'll need to raise it regularly to keep the light bulbs a few inches above the tops of the leaves as the seedlings grow.

For containers, gather recycled or new nursery pots or flats, or plastic or waxed-paper food containers. Whatever you use must be perfectly clean (wash in hot soapy water and rinse with a dilute solution of household bleach), and have one or more drainage holes punched in the bottom. For soil, the easiest approach is to buy a bag of commercial seed-starting mix made from finely milled sphagnum moss and vermiculite. Some gardeners prefer screened compost or other homemade mixes, but the commercial mixes are inexpensive and reliable. They are also sterile when

Starting from Seed CONTINUED

you buy them, which minimizes the chance that damping-off or other kinds of disease organisms will be present in the soil and attack your fragile young seedlings. Put the soil in a clean tub or pail, add warm water, and stir until it is thoroughly and evenly dampened before you use it.

There are two common methods of sowing seeds indoors. One method is to sow two or more seeds per individual pot, later thinning to leave only one seedling. Another is to sow several seeds in one container, later separating and transplanting them into individual pots. Both systems work, but sowing in one pot and transplanting later saves space and gives you an opportunity to select only the most vigorous seedlings.

If you are starting lots of seeds indoors, you will probably find it easier to sow them in nursery flats. You can sow flats in rows or broadcast the seeds over the surface as evenly as possible. Sowing in flats is a good idea when you are starting fast-growing early-season vegetables such as leaf lettuce, which can tolerate some cold and can go into the garden while there's still a chance of a late frost. You will need to thin the tiny seedlings in the flat when they become crowded. When the seedlings have grown and are ready for a second thinning, harden them off (this process is explained later in this section) and plant them out in the garden instead of transferring them to individual containers.

No matter what sort of containers you use for starting seeds, begin by filling the container about two-thirds full with damp soil mix. Carefully position the seeds on the surface, spacing them at least ½ inch apart. Cover with a thin layer of damp soil mix, vermiculite, or sphagnum moss, and pat gently to settle

To save the work of thinning later on, plant large seeds such as beans and corn one at a time, spaced at the distance the plants will need when they mature.

the seeds and covering into place. After sowing, set the containers under the lights.

Careful watering is a critical part of raising seedlings. Check them at least once a day. If the soil feels dry, looks pale or light colored, or the pot feels light when you lift it, it's time to water. The safest method is by setting each container in a shallow dish of tepid water and letting the water wick up into the soil. In less than a minute you'll see the surface darken as water reaches the top of the pot. Lift the pot to let excess water drain out, then replace it under the lights. A faster method is to water the surface with a gentle mist or spray, but be careful not to wash away the seeds or seedlings. As soon as the seeds sprout, begin using a dilute fertilizer solution instead of plain water for one watering per week.

If you sow seeds in individual pots, gently pull out or cut off the extras as soon as you can tell which plant looks best. If you sow several seeds in one container, transplant them into individual pots after they have developed their first true leaves (which follow the flaplike cotyledons, or seed leaves). You can start at one edge and lift out seedlings one at a time, using a pencil to prod and free the roots; or tip the pot and carefully slide the whole soil ball out onto your work surface, then untangle each plant from its neighbors. Hold each seedling by a leaf as you suspend its roots in the center of an empty pot, then use your other hand to fill the pot with damp soil. Don't feel obliged to transplant every seedling that grows—choose the best, and just grow as many as you need for the garden. After thinning or transplanting, put the plants back under the lights and continue watering and fertilizing regularly.

1 To transplant seedlings to their permanent home in the garden, dig a hole deep enough to position the plant at the same depth it was growing in its container.

2 Ease the plant into the hole. Support the seedling under the root ball with your trowel as you transfer the plant from pot to hole.

Vegetables to Direct-seed

Vegetables that form tap-roots are stunted by transplanting; they grow best when seeded right in the garden where they are to grow. It's better to directly sow corn; all kinds of beans and peas; root crops such as radishes, carrots, beets, turnips, and parsnips; and certain greens such as Swiss chard or spinach. If your growing season is so short that you must start some of these vegetables indoors, plant them in individual peat pots to minimize root disturbance when you plant them out in the garden.

3 Press the soil around the plant. If you live in a dry climate, make a saucerlike depression in the soil around the base of the plant to catch and hold water.

4 Immediately water the transplants thoroughly. Use a gentle spray of water to avoid dislodging the plants. Keep the soil evenly moist as the plants establish themselves.

Starting from Seed CONTINUED

TROUBLESHOOTING TIP

It's easiest to remove transplants from their individual pots if the soil is slightly dry—helping the root ball hold together, rather than crumble. But be sure to water all plants thoroughly as soon as you put them in the ground.

Seedlings need time to adjust from the protected environment indoors to the real world outdoors. Allow several days for the transition, a process called hardening off. The idea is to expose them gradually to hotter and colder temperatures, bright sun, drying winds, and overhead watering. At first, set the seedlings out for only an hour or so in the morning or afternoon, then leave them out all day, and finally leave them out day and night.

Starting indoors is recommended for seedlings of tomatoes, peppers, eggplants, and other crops that need warm temperatures to germinate and grow. Starting in a cold frame or nursery area outdoors is an option for onions, leeks, celery, lettuce, cole crops such as cabbage, and other vegetables that prefer cool temperatures. You can sow the seeds in flats or pots, or make a little raised bed and fill it with a mixture of screened compost and coarse sand. Sow and cover the seeds as usual, and make sure they don't dry out or

get frozen or overheated. Thin the seedlings to stand 1 to 3 inches apart, and water and fertilize regularly until the plants are large enough to transplant into the garden.

Whether you start them yourself, indoors or out, or purchase seedlings, the same cautions apply when you transplant them into the garden. Be careful not to snap the stems. Don't tear or crush the roots, and don't leave bare roots exposed to drying sun and air for even a few minutes—get them in the soil promptly. Water thoroughly after planting to rewet the roots and settle the soil. If possible, transplant seedlings during cloudy weather or when rain is forecast. Otherwise, plan to shade them from bright sun for a few days while they get established.

Direct sowing means sowing the seeds right in the garden, wherever you want them to grow. First, loosen the soil by tilling or digging, then use a rake to

Small seeds are difficult to sow thinly. When the seedlings come up, thin them to allow plants enough room to mature. This bed contains carrots.

Here's what happens when crops are left unthinned. These carrots are so tightly packed together that they cannot develop to their full size. Thinning would have produced a better harvest.

level and smooth the surface. If the soil is dry, water well and let it soak in for a few hours before sowing. Position the seeds evenly at the recommended spacing—be careful not to let them fall in a dense cluster. Cover with a shallow layer of soil, equal to the thickness of the seeds. Water as needed to keep the soil moist. In hot, dry, or windy weather, you can use a piece of cardboard, burlap, or a board to keep the soil from drying out, but be sure to remove the covering as soon as the seeds sprout. Use a similar covering or a barrier of thorny brush to protect directly sown seedlings from dogs, cats, birds, and children.

Usually it's a good idea to sow up to twice as many seeds as you need, then remove the extra seedlings when the plants are 1 or 2 inches tall. If the soil is damp, you can pull the seedlings out, roots and all, without damaging the remaining plants. Sometimes the seedlings can be very crowded, such as when you sow very fine seeds that are difficult to handle, or red beets, which have compound seeds, each producing three or four plants. When plants are this close together, pulling up unwanted seedlings can injure the roots of the ones you want to keep. To prevent damage, use manicure scissors to snip off the unwanted seedlings at ground level instead of pulling them up.

If you are thinning salad greens, you can thin twice, and toss the second, larger batch of thinnings into salads. Thinning may seem like a heartless job, but prompt and rigorous thinning makes way for big crops later.

Planting Methods

Crops to Plant in Hills

Hills are ideal for vining plants that like warm, well-drained soils. These include cucumbers, squash, pumpkins, melons, and gourds. For these crops, space the hills about 6 feet apart. Sow several seeds per hill and thin to the strongest two or three seedlings. In cool-climate regions, preheat the soil by covering the hill with black plastic for a week before planting. Start seedlings indoors, and set out two or three transplants per hill.

To plant climbing forms of snap, lima, or runner beans in hills, loosen the soil several inches deep in a circle 3 feet wide. Hoe the soil into a raised ring around the edge of the circle. Plant seeds 3 inches apart; later thin the seedlings to stand 6 inches apart. Make a tepee of three or more poles to support the climbing vines.

*T*here are many ways to put seeds or plants into the soil. Choosing a suitable method for each situation can make your work easier and help the plants grow better. Here are practical planting methods for some of the most popular vegetables. The step-by-step photographs on pages 68–73 illustrate several of these methods.

In most gardens, a good way to plant a row of seeds or plants is by making a shallow trench called a drill or furrow. Start by raking the ground level, then string a taut line between two short stakes to mark the row's position. Use the corner of a hoe to carve a trench under the line. Fill the row with root crops, salad greens, beans, or onion sets, leaving regular spaces between the seeds. Then use the hoe again to draw the soil back into place. Or use the trench as a guideline for setting transplants in a row—it's easier to follow a furrow than to work under a string.

Potatoes and leeks are two crops that need to grow deep in the soil. The normal practice is to pull soil up alongside the plants at regular intervals during the growing season, but this job is easier if you start by planting the seeds in the bottom of a furrow. For potatoes, make a broad furrow about 6 inches deep, then plant the seeds a few inches deep and 1 to 1½ feet apart in the bottom of the furrow. For leeks, make a furrow about 4 inches deep and set the seedlings about 1 foot apart, burying the roots and the bottom inch of the leaves. Weeks later, fill in the furrows as the plants grow. Planting in deep furrows is also a good system for corn, if the soil is warm and not too damp. (Corn seeds rot in cold, wet soil.) Cornstalks topple easily; they stand up best if you plant them in a trench and later fill in soil around the base of the stalks.

1 *For hill planting, form a mound of soil a few inches higher than the surrounding soil. Plant five to seven seeds around the top of the mound.*

In rainy climates, or wherever the soil tends to be cold and damp in spring, forming soil into ridges helps it warm up and dry out. Use a shovel or hoe to mound the soil into ridges several inches high. (Tillers or tractors do the job in large gardens.) The sides will slope naturally as the soil rolls down. The top of the soil mound can be pointed, or you can flatten it with a rake. Set transplants in place along the top of the ridge, or you can make a shallow drill to sow seeds there. Gardeners in cool climates cover ridges with black plastic to warm the soil in preparation for growing sweet potatoes, melons, or other heat-loving crops. Ridge planting is popular in dry climates, also. The roots of the plants on the ridges grow down, and you can use soaker hoses or flood irrigation to supply water directly at root level, channeling it down the furrows between ridges.

2 When the seeds have germinated and the plants are a few inches high, thin to leave the sturdiest three or four plants in each hill.

3 After the young plants have been thinned, spread a layer of straw or other loose organic mulch to conserve moisture and keep down weeds.

TROUBLESHOOTING TIP

Planting a few crops in hills is a good way to start a new garden. You don't have to dig the whole area—just prepare and plant the hills. Spread a thick mulch in between the soil mounds to smother the weeds and start improving the soil.

Certain vegetables, such as cucumbers, squash, and melons, are often planted in hills, or low mounds of soil. Hills have all the benefits of ridges, but are easier to form with hand tools. For each hill, loosen the soil in a circle about 3 feet wide and 1 foot deep. Then pull soil in from the edge to make a flat-topped cone in the center, about 2 feet wide and a few inches high. The soil mounded in the cone gets warmer and drier than the surrounding soil—this gives warm-weather crops a head start in cool climates. Meanwhile, the doughnut-shaped trench around the edge catches rainfall or irrigation water and directs it efficiently to the roots. No wonder gardeners have been planting in hills for thousands of years!

Mulch around seedlings to help retain soil moisture between waterings. These hill-planted squash seedlings are ready for thinning.

Planting Methods CONTINUED

EARTH•WISE TIP

Climbing vines get quite heavy and need a sturdy support. If you use any kind of twine, make sure it's thick and strong. But use natural fiber, not synthetic, twine, so you can toss the whole tangle of vines and twine onto the compost pile.

1 When planting pole beans or other climbers, set up the supports for the vines before you plant seeds. Make a furrow of the appropriate depth with a trowel or hoe.

2 Sow the seeds along the bottom of the furrow. To save time where the frost-free growing season is brief, you can pregerminate seeds indoors between moist paper towels.

1 If a tomato seedling has grown leggy before you get it into the garden, try planting it on its side in a shallow trench. Additional roots will grow from the buried stem.

2 Lay the plant in the trench, and cover the stem to the lowest set of leaves. Firm the soil over the buried stem. The top of the plant will straighten out as it grows. Water well.

3 Cover the seeds with soil to the recommended depth, firm the soil, and water thoroughly. Then water as needed to keep the soil evenly moist until seedlings appear.

4 When plants are a few inches high, thin them to the correct spacing. Plants crowded too close together will not produce as well as plants given sufficient room to mature.

3 To exclude weeds and to capture additional heat for plants in cool-climate gardens, fit a "collar" of black plastic mulch around the base of each plant.

4 Leave space between the plastic and the stem to permit watering. Cover the plastic with soil to hold it in place throughout the growing season.

Planting Methods CONTINUED

When asparagus spears push through the ground in early spring, harvest by cutting them off at ground level with a knife or clippers.

After the asparagus harvest is over, allow the tall, feathery foliage to develop on the remaining unharvested spears. The plants make an attractive edging or divider in the garden.

Planting asparagus is a special project, because unlike most vegetables, which are short-lived annuals, asparagus is a perennial that can produce for years. Think carefully about where you want to plant asparagus. You don't have to put it with your other vegetables; instead, you might use it as a hedge or backdrop for a flower bed—it's a handsome plant that makes a dense mass of feathery foliage, rich green all summer and golden yellow in fall (see the photo above). In any case, it needs full sun and rich, well-drained soil.

You can start asparagus from seed and raise the seedlings in a nursery bed for one season before transplanting them to the permanent bed, or purchase dormant crowns at garden centers or from mail-order catalogues for early spring planting. Either way, follow the step-by-step photographs on the opposite page for planting asparagus.

A few other vegetables—artichokes, Jerusalem artichokes, and rhubarb—are also perennials that benefit from careful site selection and generous soil preparation. Spend the time and effort to get these vegetables off to a good start, and you'll enjoy the harvest for years to come.

1 Asparagus plants make an attractive edging along a path. To plant dormant crowns, first dig a trench 6 inches deep in most soils (8 inches in sandy soil).

2 Set the crowns along the bottom of the trench, 1½ to 2 feet apart. Place the eyes (growth buds) facing upward, and spread out the roots. Cover with 2 inches of soil.

3 As the plants grow, periodically add 1 to 2 more inches of soil at a time, until the soil in the trench is level with the surrounding soil.

4 When the trench is filled with soil, mulch around the ferny plants with compost or shredded leaves to retain moisture and discourage weeds.

Care after Planting

After the rush of the planting season has subsided, but before you forget the details, take time to make notes of what you've planted. Mark planting dates on a calendar, sketch a map to show what's planted where, and write comments in your garden notebook. Taking a little time now will save you lots of time later, and you will know exactly what's planted where and which varieties are doing well or poorly.

Vegetable plants are most vulnerable during the first few weeks after you put seeds or transplants in the ground. This is the time when you must take special care to protect them from stress and damage.

A major responsibility is to keep them from drying out. Watch seedbeds closely. The surface of the soil can dry out fast on sunny or windy days, and at this stage one severe drying out can put an end to your crop. Watch transplants, too—their roots aren't deep or extensive enough yet to keep up with water lost through the leaves. Frequent light sprinklings—as often as once a day—may be required, especially for crops planted during the summer.

Too much water can be as harmful as too little, particularly if it comes as a cloudburst. Check the garden after a downpour. Sprinkle a fresh covering of soil over any seeds that have been washed bare. Use a springy rake or pronged cultivator to gently prick and loosen the soil over newly sown seeds if heavy rain has packed it down, so that the seedlings won't have to struggle through the crust that forms when packed soil dries. Loosen and lift the leaves of any seedlings or transplants that have been pressed down flat against the soil.

Young plants are usually sensitive to extreme cold or heat. In spring, pay attention to the weather forecast for the first week or two after transplanting tender seedlings such as tomatoes. If frost is predicted, be sure to protect the seedlings with floating row covers, cloches, cardboard boxes, or other coverings. In summer, erect covers or shades of some kind to protect new transplants from the hot sun for several days.

Pests, large and small, are especially threatening to tender young plants. Crows, starlings, and house sparrows all pluck at seedlings. Try deterring them with rubber snakes, inflatable owls, or shiny foil

As seedlings, cabbage and its relatives—broccoli, cauliflower, Brussels sprouts, kale, collards, and turnips—are at risk of attack by cutworms. Protecting them with cutworm collars can enable them to mature and produce their crop.

1 *If you have problems with cutworms, protect the base of seedlings with cardboard collars. Cut 4-inch-wide strips from paper cups or empty milk cartons.*

2 *Place a collar around each seedling, pushing the collar 2 inches into the soil. The collar should fit loosely around the stem, to allow for growth.*

ornaments that move in the breeze. Deer, woodchucks, rabbits, gophers, and other critters can eliminate a row of seedlings in a single swallow. Blood meal, human hair, mothballs, and various pungent herbs have all been recommended as repellents, but fences and traps are the best strategy in serious cases. Wild animals aren't the only problem; you may need some kind of barrier to keep out running dogs or digging cats.

One of the best new garden products is a lightweight spunbonded polyester fabric called a floating row cover. You can spread a layer of this fabric directly over the plants or use wire or plastic hoops to support it like a tent. The fabric is translucent and porous, so sunlight, air, and water pass freely, but a row cover can protect young plants from dry wind, hot sun, chilly nights, hungry birds and beasts, and—

best of all—flying insects. Spread the cover loosely over a row or bed, and secure all the edges by weighting them down with rocks, boards, or soil. This foils flea beetles, onion flies, cucumber beetles, leaf miners, and many other insect pests. (Remove the covers when plants bloom, to allow pollination. Then you can replace them.)

Row covers don't stop cutworms—fat caterpillars that hide in the soil by day and come out at night to bite through the stems of tender young plants, toppling them much as a beaver fells a tree. If cutworms are a problem in your garden, you can protect individual transplants by putting collars or disks of heavy paper or cardboard around the base of the stems. The photographs above illustrate how to make a cardboard collar.

Weeding and Cultivation

TIMESAVING TIP

To keep weeding chores from becoming overwhelming, try to spend just 10 minutes a day weeding the garden. You'll be surprised how pulling only a few weeds each day will enable you to keep them from getting out of hand. Being in the garden on a daily basis will also let you spot pest and disease problems while they are minor and easy to manage.

A sunny site with loose, fertile, moist soil is the ideal situation for growing vegetables. Unfortunately, it's also the ideal spot for weeds. That's a problem, because weeds compete with your crops, reducing yields. They can also harbor pests and diseases.

There are various approaches to controlling weeds, depending partly on what kinds of weeds you have. Basically, weeds can be divided into two categories: those that die when you cut them off below ground with a hoe or pull them out, and those that sprout back manyfold like the sorcerer's apprentice.

Weeds in the first category are mostly annuals such as amaranth, lamb's-quarters, chickweed, ragweed, purslane, galinsoga, and crabgrass. A two-step approach can give you a head start on controlling them. First, plan ahead and till or cultivate the soil a week or two before you want to plant, and water well to encourage the weed seeds to sprout. Then hoe lightly to uproot and destroy that first crop of weeds before you sow seeds or set out transplants. As the season continues, keep up with the weeds by hand-pulling them, cutting them off with a sharp hoe or hand-weeding tool, or running a cultivator between the rows. This is important: don't let annual weeds mature and go to seed! One overlooked weed can sow thousands of seeds that will haunt you in future years.

Perennial weeds such as ground ivy, bindweed, Canada thistle, Bermuda grass, and quack grass cause a lot of anguish. No matter how many times you pull or hoe out the shoots, more keep growing back.

It takes heroic persistence to kill these weeds by plucking off their tops. Cultivating or tilling only makes the problem worse, because it breaks the underground runners into tiny bits, each of which makes a new plant. One way to control such weeds is to suspend gardening for a season and bake the soil by a process called solarization. Till the plot and water well, then cover it with a large sheet of clear plastic and let it bake in the summer sun for four to eight weeks. The soil gets hot enough under the plastic to kill most weed runners and roots.

Another method of weed control is to starve the perennial weeds by keeping them in the dark. Plant the garden in rows, and line the space between rows and a 4-foot margin around the edge of the garden with several sheets of newspaper, topped with several inches of hay, straw, leaves, clippings, or other mulch. Pluck any stray shoots that do make it up through the mulch, and leave the mulch in place for an entire growing season.

Cultivating can help eliminate weeds, but its primary function is to loosen the surface of the soil, breaking through any crust and allowing water and air to penetrate more easily. It's most useful in heavy clay soils, and unnecessary in sandy soils. Be careful not to disturb any roots by working too deep or too close to the plants.

There are dozens of styles of weeding tools for different tasks. If you have a large garden planted in rows, using a wheel hoe, push-style cultivator, or small rotary tiller is the fastest way to weed between the rows. Then follow with a hoe or hand weeder to eliminate weeds within the rows.

If your garden is not mulched, it's important to keep up with the weeding. Weeds are easiest to pull when they are small, like the ones in this garden.

Hand-weeding tools are sufficient for a small garden, especially if the paths are covered with a weed-suppressing mulch. Cultivator-type weeders usually have three tines that loosen the soil and lift up weeds, roots and all. Use one to lift crowds of weed seedlings while they are still small. Work on a sunny day so that the seedlings will quickly dehydrate. Slicer-type weeders have strong but thin blades with a sharp cutting edge. Use a slicing tool to cut individual weeds just below the surface of the soil. Sharpen the cutting edge regularly with a metal file. A sharp tool is easier to use and does a better job. If you shop around and check mail-order tool catalogues, you'll find both cultivator- and slicer-type weeders with either long or short handles, so you can work standing up or kneeling down.

Mulching

Mulching offers many benefits, both for the plants and for the gardener. Mulch can control weeds, conserve soil moisture, and regulate soil temperature. Mulched plants grow better and yield more, and a mulched garden is easier to care for.

Many different natural materials and manufactured products can be used for mulching. The natural materials include straw, hay, grass clippings, shredded tree leaves, pine needles, chipped brush or bark, sawdust, and peanut or cocoa bean hulls. Any of these materials can be used fresh, but if you have the time and space, it's better to make a pile and age or compost them for a few weeks or months before spreading

them as mulch. These materials decompose over the course of the growing season and can later be worked into the soil, adding valuable organic matter.

Manufactured products used as mulch include newspaper, brown paper bags, corrugated cardboard, rolls of paper or synthetic landscape fabric, and rolls of clear or black plastic film. Paper-based products break down and can be worked into the soil at the end of the growing season. Plastic products eventually break down, too, if exposed to the sun, but have to be removed and discarded each year.

At the beginning of the gardening season, spread mulch on the paths and around the edges of the garden to control weeds there. Paper products, coarse hay, or bark chips are all good for this purpose. If

Mulching keeps the garden neat, cuts down on weeds, and conserves soil moisture. Spread a coarse mulch, such as straw or dry leaves, 3 or more inches deep.

1 *Straw makes a good mulch for vegetable gardens and adds organic matter to the soil when it decomposes. It is less likely than hay to contain weed or grass seeds.*

2 *Spread the straw around and between plants in a layer 8 to 10 inches deep; it will compress over the course of the growing season. If the mulch becomes thin, add more.*

you have permanent paths, consider investing in landscape fabric topped with bark chips, a combination that looks good and lasts for years. As you plant the garden, you can use a thin sprinkling of grass clippings or pine needles (along with the regular covering of soil) to keep newly sown seedbeds moist. Seedlings can grow right up through a light mulch. Later, spread a thick layer of straw, leaves, hay, or hulls around and between larger seedlings and transplants.

Generous mulching greatly reduces the amount of time you'll have to spend weeding and watering a garden. However, there are problems associated with organic mulch, especially in rainy climates. Damp mulch can harbor fungus diseases that infect plant stems and leaves, and it hosts slugs, snails, and earwigs. In rural areas especially, mulch provides a haven for mice, rats, and voles, whose nibbling can damage or destroy a crop.

Plastic mulch has serious drawbacks—it's relatively expensive, it's hard to reuse or recycle, and (to many gardeners) it looks ugly. But plastic is very useful for preheating the soil in spring; this makes it possible to grow melons, sweet potatoes, and other heat-loving crops in northern or high-altitude gardens. Prepare the soil and cover it with plastic for a week or two before planting. Clear plastic does a faster job of warming the soil, but doesn't inhibit weed growth. Black plastic warms the soil and also controls weeds.

Fertilizing

Plants can absorb nutrients through their leaves, as well as through their roots. Spraying or sprinkling a dilute fertilizer solution on the foliage—called foliar feeding—is a good way to boost new transplants or young seedlings that don't have big root systems yet.

*I*n addition to sun, air, and water, plants need nutrients to grow. These include nitrogen, phosphorus, potassium, calcium, magnesium, iron, and other elements. Although fertile topsoil contains many nutrients, the ordinary soil found on most potential garden sites doesn't supply enough nutrients for maximum growth and yield. Carefully measured and well-timed applications of organic or inorganic fertilizers give a real boost to a garden. Fertilized plants look better, grow faster, get bigger, and yield more vegetables.

A major question is how much fertilizer to apply to the garden. The answer varies, depending on your soil. Truck farmers, market gardeners, and other serious vegetable growers don't guess about this—they have soil samples tested annually and fertilize according to the test results, adding only the nutrients that are required. A soil test gives you very helpful information, and most states offer inexpensive soil tests through the agricultural university or county cooperative extension service. Without soil test results, you're taking a chance when you fertilize. If you don't use enough, your plants won't achieve their full potential. However, adding too much fertilizer wastes your money and can harm your plants and pollute the local watershed.

Any garden center sells bags of inorganic fertilizer. They are labeled with three numbers that indicate the percent, by weight, of nitrogen, phosphorus, and potassium (in that order)—the three nutrients that plants need in greatest amounts. For vegetable gardens, fertilizers such as 5-5-5, 5-10-5, and 10-10-10 are commonly recommended and generally suitable. Following the recommendations that accompany your soil test results, you can apply measured amounts of inorganic fertilizer before planting the garden.

Sprinkle it evenly across the soil, and till or rake it in well. Give a second, smaller application (called side-dressing) in midseason to crops that take more than two months to mature. Sprinkle it along the rows or around the base of a plant, scratch it in with a cultivator, and water well.

Organic fertilizers come in several forms. Fresh, aged, composted, or dried animal manures are the most commonly available. Manure is cheap by the truckload in farming regions, but more expensive in packaged form. Manure supplies moderate amounts of the major nutrients, along with a wide range of trace elements. In addition to common steer, cow, horse, sheep, and poultry manures, there are exotic offerings, such as zoo animal manures, bat guano, cricket droppings, and earthworm castings. Use whatever you can get and afford. It's a good idea to combine fresh manure with hay, grass clippings, garden waste, or leaves in the compost pile for faster and hotter composting, then use the mixed compost as a soil amendment or topdressing. Or you can spread aged manure directly on the soil and till it in well before planting. Aged or composted manures are not very strong, and you're not likely to overapply them unless you're trying to dispose of the waste from your own animals. If you have to buy manure, you probably won't use too much.

Blood meal, feather meal, fish meal, and other animal by-products are sometimes dried and bagged for use as manure. Prices vary widely. All are good, but smelly, fertilizers. Cottonseed meal, seaweed extracts, alfalfa pellets, and other plant products and by-products are also good organic fertilizers that supply a range of vital nutrients.

If you use granular fertilizers, side-dress rows of vegetables several times during the growing season. Pull aside the mulch, scatter fertilizer evenly alongside the plants in the row, scratch it lightly into the soil, water, and replace the mulch. Netting in this garden protects beans from being eaten by birds.

Staking and Training

It's hard to secure a plant to a smooth, slippery metal or wood pole. With nothing for the ties to catch onto, they slide down the pole. For better support, use rough-cut lumber, saplings with bark and limb stubs, or bamboo.

*T*here are advantages to training vining plants such as tomatoes, cucumbers, melons, beans, and peas to grow up stakes, tepees, or trellises. It saves space to train plants vertically instead of letting them sprawl over the ground. Raising the vines up into the air means the developing fruits are surrounded with warm, dry air instead of cool, damp air, so they ripen faster and are less susceptible to rotting.

To train individual tomatoes or other plants, drive 6-foot or longer stakes at least 1 foot into the ground. Set the stakes in place before you sow or plant so that you won't have to disturb the plant's roots later. Use strips of soft cloth, old nylons, or soft twine to tie the growing stem loosely to the stake. Check weekly, adding new ties every 6 inches or so.

Cucumbers and melons can't climb a smooth stake, but their tendrils will grip a net, trellis, or wire-mesh fence enough to hold up the vines. It's okay to let cucumbers dangle from the vine, but it's a good idea to support heavy melons with a cloth sling, similar to the one the legendary stork uses to deliver babies. Wrap it under the melon, and tie the top to the trellis.

Peas are weak climbers and may need to be pushed against a fence or tucked into the squares of a net. If the vines slump, stitch them to the support by threading a ball of twine back and forth and pulling it tight enough to hold up the vines.

Pole beans climb readily, twining around any pole, stake, or other support. Just bring any stray runners back close to the pole and tuck them under another stem, and they'll soon start twining again.

Nasturtiums will climb if you fasten them to a trellis or other support, and their bright flowers add a shot of edible color to the garden.

1 *Tomato plants produce nonfruiting suckers that create bushy growth but channel some of the plants' energy away from fruit production.*

2 *The suckers are leafy stems that develop in the axils of the main stem and side branches, and produce no flowers. Pinch or clip off a sucker where it joins the stem.*

3 *Removing suckers allows the plant to concentrate on producing fruit. If the sun is intense, however, the tomatoes could suffer sunburn if too many suckers are removed.*

4 *As the plant grows, the stems will need support, especially in indeterminate tomato varieties. If using stakes, attach the stems with strips of cloth or soft yarn.*

Special Care

TROUBLESHOOTING TIP

Protect developing squash and melons from soil-borne diseases and insect pests by perching them on a flowerpot, tin can, brick, or block of wood. Up in the air, away from the cool damp ground, they ripen faster and develop better color and flavor.

Most vegetables grow readily without much attention, but sometimes a little special care pays off in increased quality or yields. The secret is knowing what to do and when to do it.

Blanching is a process of denying light to certain vegetables in order to make them more tender and (in some cases) less bitter, and more attractive. It means different things for different plants. Blanching cauliflower means pulling the leaves over the top of the developing head. If exposed to the sun, cauliflower turns a dirty cream color instead of snowy white. In contrast, to blanch leeks or celery, pull several inches of soil around the base of the plants so that they fade from green to pale yellow or white. To blanch chicory or endive, cover the dormant roots (often potted and stored in a cool cellar) with an inverted pot or box; as a result, the new shoots will emerge in the dark and stay tender and pale.

In a practice similar to blanching, you can cover the lower parts of potato stems with several inches of soil or mulch (this is known as hilling). The potatoes form along the buried stem and must develop in the dark. If potatoes are exposed to the sun, the skin turns green and produces a poisonous compound.

The opposite of blanching is letting more light reach developing parts of the plant. Push onion tops over sideways so that sun can shine on the swelling bulbs. In cool or cloudy climates, remove corn tassels after the ears have set, to let more light reach the leaves. Pruning extra foliage off leafy tomato vines or pepper plants can help those vegetables ripen, but be careful not to remove too many leaves at once—sudden exposure can cause sunburn.

Although some plants keep growing in late summer and fall, where winter comes early, there isn't time for late-formed vegetables to mature. In this case, it's bet-

1 *To produce tight, white cauliflower heads, you need to blanch them. Unblanched heads turn brownish and the curds grow loose and "ricey," although flavor is unaffected.*

ter to force the plant to put all its energies into what's already started. A few weeks before the expected fall frost, cut the ends off pumpkin and winter squash vines, and the tops off tomato plants. At this point it's too late for tiny green fruits to develop and ripen. A few weeks after the first hard frost, cut the tops off Brussels sprouts plants to stop further growth.

Sometimes you need to help with pollination. In small corn patches, shake the stalks daily to help the pollen reach the silks. Shaking tomato plants will self-pollinate individual flowers and initiate fruit set. If you're growing squash, cucumbers, or melons under row covers to keep out insect pests, you'll have to lift the covers daily and hand-pollinate with a clean paintbrush, transferring pollen from male to female flowers.

2 Draw some of the plant's large outer leaves loosely together over the head to completely enclose the head and block out light.

3 Secure the leaves with a rubber band or piece of string. Keep the head covered until it is ready to harvest (open the leaves periodically to check its development).

1 If you grow squash under floating row covers, remove the covers while plants are in bloom, or pollinate them by hand. Transfer ripe pollen from male to female flowers.

2 Transfer the pollen with a soft, clean artist's paintbrush, or remove the pollen-bearing flower and touch its anthers to the sticky pistil in the center of the female flower.

Controlling Pests and Diseases

*I*t's not surprising that insect pests are attracted to vegetables; after all, they thrive on a diet of succulent, nourishing food just as we do. What is surprising, though, and aggravating to a gardener, is just how much damage these tiny pests can do. Insects may be small, but they're numerous, and they have big appetites.

However, not all insects are villains, and the first step in pest control is identifying the problem. Just because you see an insect on a leaf doesn't mean that insect is harming the plant. It may just be resting there, or it may be a pollinator, or it may even be a predator that eats other insects. Don't knock it off until you take a closer look. Can you see any damage? Are there any unusual chew marks, holes, dots, or specks on the leaf? Look on other leaves, and on adjacent plants. Do you see more of the same insects? Don't forget to check the undersides of leaves, the leaf axils (where leaf joins stem), and the young growing tips when hunting for pests. If you don't see damage,

and the insects aren't abundant, you may not have a problem, but it's a good idea to check again every day or so to see if the situation changes.

If you do see damage, but you can't identify the insect pest, try to catch it in a glass jar. Put a few of the damaged leaves in, too, and take it to an experienced gardener, a garden center, a county cooperative extension service agent, or some other authority for identification. Chances are that the same kind of pest is attacking other gardens in your community, and an authority will quickly recognize it.

Identification isn't really difficult, because only a limited number of pests attack any particular crop, and each does characteristic damage. If you find tiny scattered holes in your eggplant leaves, for example, it's probably an infestation of flea beetles (tiny black beetles that hop when disturbed). With a few years' experience, you can learn to identify most of the insect pests that are likely to visit your garden.

The praying mantis is a voracious eater of insects, attacking aphids, beetles, flies, wasps, and caterpillars.

Ladybugs are warriors in the garden. Both larvae and adults consume aphids and other pests.

The lacewings that hatch from these dangling eggs will devour the aphids that will emerge from the orange eggs on a nearby plant.

Parasitic wasps lay their eggs on host caterpillars, such as the tomato hornworm. When the young hatch, they prey on the unfortunate host.

1 *Floating row covers can protect cabbage from being damaged by pests. Lay covers over the plants loosely enough so that the plants have room to grow.*

2 *When the heads are ready to harvest, roll back the covers and save them for reuse. These lightweight row covers also afford excellent protection for new transplants.*

EARTH•WISE TIP

You can trap slugs and pill-bugs under boards, inverted flowerpots, or grapefruit rinds. They seek shelter at dawn and will congregate in the traps, where you can easily find them during the day. Drown them in a can of salty water.

Learning to identify insect pests is the first step toward controlling them. The next step is choosing a suitable treatment. If they're relatively big and slow, and there aren't too many, you can pick them off by hand and drown them in a can of salty water (add a tablespoon of salt to a cup of water).

Spraying is an easier treatment if the insects are too numerous, too quick, or too small to catch. You don't need an arsenal of strong chemicals. Most garden centers today offer earth-friendly alternatives, such as insecticidal soap; pyrethrum, rotenone, and other plant-derived compounds; and *Bacillus thuringiensis* (Bt), a bacterium that kills many kinds of leaf-eating caterpillars but doesn't harm other insects. Mail-order suppliers offer these plus many other specialized pest-control products.

Whatever the pesticide, take time to study the label before applying it, and be sure to follow directions carefully. Insecticidal sprays—even organic ones like insecticidal soap and pyrethrum—must be handled with care. The great advantage to so-called organic insecticides is that they break down quickly after application and do not linger in the environment. But they are toxic when you apply them, so wear gloves. Bear in mind, too, that most insecticides kill beneficial insects as well as pests. Although convenient, they are really best used as a last resort.

Some gardeners swear by homemade sprays that use small amounts of common household ingredients such as rubbing alcohol, liquid dishwashing detergent, salad oil, ground chili peppers, or crushed garlic mixed with water. You won't do any harm by experimenting with these sprays, and they might work well

Controlling Pests and Diseases CONTINUED

To avoid this wireworm damage on carrots, trap wireworms with pieces of potato placed in the garden.

Cabbage loopers, green caterpillars that turn into butterflies, chew ragged holes in the leaves of cabbage-family members.

Control corn earworms with a drop of mineral oil inside the tip of each ear after the silk has wilted.

Tomato hornworm attacks peppers and eggplants as well as tomatoes. Handpick and destroy the caterpillars.

for you. Just be sure to thoroughly wet both the tops and the bottoms of the leaves. Check the plant a day after spraying to confirm that the treatment was effective, and repeat at weekly intervals if needed.

Sometimes prevention is the best control, and the easiest way to prevent many insect problems is to cover the plants with lightweight row-covering fabric. Usually, it's best to put the fabric in place right away, as soon as you sow seeds or set out transplants, but you'll probably have to loosen it later as the plants grow. Be sure to fasten down the edges, so that pests can't sneak inside.

Row covers (or garden blankets, as they are sometimes called) are especially effective at guarding crops from flying insects that lay eggs, which later hatch into caterpillars or grubs that feed on the plants. The notorious borers that can demolish a promising crop of zucchini are one example of this type of pest. The covers prevent the insects from laying their eggs on or near the target plants, and the harvest is spared. Since light, air, and water readily penetrate the lightweight fabric used for row covers, many gardeners leave the covers in place throughout the growing season. You must, however, remove the covers when plants are blooming, to allow pollination, or pollinate the plants yourself by transferring the ripe pollen from flower to flower with a clean paintbrush.

Some pests hide for the winter in dead stems or other plant litter, or just under the surface of the soil. For that reason, it's a good idea to remove all crop debris from the garden and hot-compost it to kill most of the insect eggs and larvae. Many gardeners also till or fork over their garden at the end of the season, exposing buried insects to birds and frost.

Removing plant litter is also important in disease control. Vegetable plants are susceptible to many

Vegetables Resistant to Disease
Vegetable varieties differ in their susceptibility to common diseases; resistant varieties continue to grow and produce in conditions where regular plants would suffer and die. Look for information about disease resistance in seed-catalogue descriptions. The disease-resistant tomato variety shown here, 'Ultra Girl', ripens well in northern gardens.

kinds of diseases. Various blights, molds, and mildews cause wet or dry blotches and spots on the stems, leaves, flowers, and fruits. There are few effective controls, but several strategies of prevention. One step is destroying infected plants by burning them or sending them to the landfill. (Composting infected plants risks spreading, rather than eliminating, the disease.)

Keeping the leaves dry is often recommended as a way of preventing fungal diseases, but that's impossible if you live in a humid or rainy climate. The best you can do is choose a site with full sun, space plants extra-far apart, and hold back on moisture-retentive mulch so that leaves wet by dew or water will dry as soon as possible.

Don't let pest and disease problems get you down. If you're a new gardener, ask around the community for advice. In any region there are some crops that are easy to grow and others that are magnets for trouble. Learn which are which, plant accordingly, then relax and enjoy much easier gardening.

Harvesting

*A*fter weeks or months of watching your vegetable plants grow, there comes a point of decision: when should you start harvesting? The answer varies from crop to crop.

Some vegetables mature so quickly that you need to pick daily to catch them at their prime. Asparagus is like this, and so is okra. If you don't pick new asparagus shoots soon after they emerge, they quickly grow tough and woody. Harvest okra pods two or three days after the blossoms drop off, while they're still soft and fuzzy. If you wait more than a few days, they get so tough that cooking won't soften them.

Watch broccoli and cauliflower heads closely. They enlarge rapidly over a week or two, then seem to stabilize—that's when to pick them. Don't wait too long, especially in warm weather (fall crops develop more slowly). If you do, the heads get loose and chaffy and produce dozens of little yellow flowers that are edible, but not very tasty. It doesn't matter if you have more broccoli or cauliflower than you can eat; pick it anyway, as soon as it is ready. You can store it in the refrigerator for a few weeks, blanch and freeze it, or give it away. Likewise, pick and refrigerate radishes and kohlrabi as soon as they reach full size. If you wait even a few days too long, they get too tough to eat.

Cucumbers and summer squash have the ideal combination of rich flavor, tender skin, and still-undeveloped seeds if picked before the flowers have shriveled and fallen off. Of course they're still edible later, but eventually they turn into bland blimps with thick skin and tough seeds. If you can't eat them all, keep picking anyway and give them away, and make a note to start fewer plants next year.

Sweet corn takes only seven to ten days to traverse the span from not quite ready (when the kernels are still small and watery), to over the hill (when the kernels are swollen and chewy). The traditional test to tell when sweet corn is ready to pick is to puncture a kernel with your fingernail. If a milky liquid squirts out, it's time to harvest; if the kernel does not contain any liquid, the corn is past its peak.

Snap beans are most succulent and tender when the pods are still thin and straight as a pencil. A week later, when the swelling seeds are evident, the pods lose their flavor and develop a hard, papery texture.

Sugar snap peas, which are edible at any stage from flat pods to mature peas, are best if you pick them when the pods are fully grown and the peas inside are plump. The pods will still be juicy, sweet, and succulent. Snow peas, on the other hand, are best harvested while the pods are still flat. When the seeds inside develop, the pods become tough and starchy. English peas and lima beans are exquisitely tender (though not particularly tasty) if picked when the seeds are still small and only half-fill the pods, but it takes a big panful of pods to yield a single serving at that stage. There's almost no sacrifice in tenderness, and a big gain in yield, if you wait a few days until the seeds come together and fill the pod. (Note: If you want nutritional value, rather than tenderness, it's better to let all kinds of peas and beans ripen on the vine until the pods are tan and dry—mature peas and beans are filled with starch and protein, not just sugar water.)

It's wonderful to bite into a rich vine-ripened tomato, still warm from the sun, but in fact picking tomatoes doesn't require very careful timing. You can pick them any time after the color changes from green to yellow-orange and let them ripen on a warm windowsill, and the flavor will be just as good. If you

The bounty of the garden, in its infinitely rich variety, makes all the work well worthwhile.

raise quantities of tomatoes for canning, simplify harvesting by going through the patch every few days and picking all that are colored, then spread them on a porch table for a few days to finish ripening. Unfortunately, commercial growers have taken this to an extreme, but the truth is it's much easier to pick, carry, and sort tomatoes that aren't squishy ripe.

Bell peppers aren't fussy; if you don't pick them green, they just get prettier and sweeter as they ripen to red or yellow. Chili peppers will turn red and shiny and hang on like Christmas decorations until you get around to picking them.

Leaf lettuce, spinach, New Zealand spinach, kale, collards, Swiss chard, and other greens used for salads and cooking usually last for weeks or even months in the garden, given plenty of water and not-too-hot

Harvesting CONTINUED

1 *Harvest new potatoes when they grow large enough, anytime after plants stop flowering. When plants begin to die down in late summer, dig up all remaining tubers.*

2 *Handle the potatoes carefully. Save very small tubers for planting next year. Let potatoes dry for several hours, brush away the soil, and store them in a cool, dry place.*

weather. Pick all you want, as often as you want, by harvesting leaves from around the edge or the bottom of a plant—more will grow up in the center.

You can start gathering tender "new" potatoes when they reach the size of walnuts (usually a few weeks after the vines start flowering). To harvest potatoes for winter storage, though, wait until the tops have turned brown and withered. Similarly, you can start pulling carrots when they're as big as a finger and leeks when they're as thick as a hot dog. Or you can leave them in the ground all fall (all winter, too, if you mulch to protect them from freezing), to harvest whenever you wish.

If you grow onions, you can start harvesting them for scallions several weeks after planting sets or seedlings (they take a bit longer from seed). In fact, you can harvest seedlings to thin the plants and allow space for the other bulbs to develop to maturity. Let onions intended for storage grow all season until the top growth falls over and dies. When the leaves die back, let the onions remain in the ground for a week or so to toughen the skins. After the leaves are completely dry and shriveled, dig up the bulbs. Let them air-dry (outdoors if the weather is nice) for a few days to a week. Then bring the onions indoors and let them dry for another month before storing them.

Collecting and Saving Seeds

Are Hybrids Better?

Most of the varieties offered by modern seed companies are hybrids made by controlled breeding between two carefully selected parents. These varieties are usually labeled on the packet and in the seed catalogue as "hybrid" or "F1 hybrid." The seeds cost more, but hybrids often have increased vigor, better disease resistance, and higher yields. Hybrid crops tend to mature uniformly, reaching the same size at the same time.

Saving seeds from hybrid plants is often disappointing. The next generation may not have as much vigor, productivity, and overall quality as their parents. If you want to make a regular practice of saving your own seeds, it's best to start with varieties marked "heirloom" or "open-pollinated." These varieties come true from year to year and often have outstanding flavor.

What did gardeners do before there were seed companies? They saved their own seeds from year to year, and you can, too, at least for some crops.

Start with peas and beans. They generally come true from seed; that is, the seedlings resemble their parents. Choose healthy plants from a variety that grows well for you, let some pods ripen on the vine until they are dry, then shell out the seeds.

Save seeds of lettuce, radishes, beets, turnips, and mustard greens by letting a few plants go to seed, then gathering the seedpods just when they turn yellow-brown. Put them in a paper bag indoors to catch the seeds as they dry. Use the same technique to save seeds of broccoli, cauliflower, or kohlrabi.

Insects can cross-pollinate different varieties of tomatoes, peppers, eggplants, or okra, but the seeds will come true if you grow only one variety, or if you isolate one plant at a distance from related kinds and save just its seeds. Likewise, insects are likely to cross different kinds of cucumbers, melons, squash, or pumpkins, and the offspring of uncontrolled crosses are usually motley. To save these seeds, you need to exclude insects and pollinate by hand.

Corn pollen is easily carried by the wind and can float several hundred yards from one variety to another. However, if you have the only corn patch in the neighborhood and grow varieties that tassel at different times, you can save the seeds. Choose well-filled ears and let them ripen on the stalks until the husks are dry and tan before picking.

Familiar vegetables look different when they go to seed. Clockwise from top left are broccoli, parsnips and beets, carrots, lettuce, and cabbage.

Extending the Season

Where winters are mild, all along the Pacific Coast and across the southern United States, gardeners can easily pick vegetables all year. There are many crops that can endure occasional frosty mornings. Although growth rates are slower in the short days of winter, the garden keeps producing. These gardeners enjoy an ever-changing menu of seasonal crops, but there is one drawback: they never get a week off. Where the season runs year-round, the chores never stop.

It's another story in cold-winter climates. There, the growing season never seems long enough to accommodate everything you want to grow, and the winter diet of stored and preserved vegetables really hones your appetite for anything picked fresh. Fortunately, there are many strategies a gardener can use to extend both ends of the growing season, in spring and fall.

You can get a head start in spring, planting weeks before your neighbors do, by taking steps to warm and dry the soil and by protecting young plants from frosty air. Especially when the weather has been cool and wet, you might try solarizing the soil in part of the garden. Stretch a sheet of clear plastic over the soil and bury or weight down the edges. With a reasonable amount of sun, in a few weeks the soil under the plastic should be warmer and drier and, if you are lucky, ready for planting with early crops. An added bonus of solarizing is that the heat kills weed seeds as well as drying the soil.

One of the easiest and best ways to get a head start in spring is by using a cold frame. A cold frame can be as simple or as fancy as you want—the truth is, they all work. What's important is full (all-day) expo-

1 *A cold frame allows you to start seedlings early in spring. When the temperature gets above 40°F, open the lid partway to keep the plants from overheating.*

sure to the sun and some easy way to adjust the lid, opening it for ventilation on warm, sunny days and closing it at night or on chilly, cloudy days. Unless you're home during the day, an automatic vent opener, available from mail-order garden suppliers, is the best cold-frame accessory you can buy.

There are two ways to use a cold frame as a nursery for raising seedlings to transplant around the garden. You can use the frame to preheat and protect a bed of well-prepared soil, and sow the seeds directly in that nursery bed. Or you can raise seedlings in pots or flats that you put in the cold frame. Potted seedlings require more frequent watering than bed-raised seedlings, but suffer less shock at transplanting time. You can start seedlings in the cold frame 6 to 10 weeks before the last frost.

2 *You can also harden off seedlings in a cold frame. Open the lid a bit more each day over the course of a week or two, until the plants are ready to go into the garden.*

3 *To get an early crop of tomatoes, plant a seedling in the cold frame several weeks before it could safely go into the garden. When the weather warms, leave the frame open.*

In the past gardeners made hot beds by digging a bed about 2 feet deep, removing the soil and replacing it with fresh stable manure, then adding a layer of soil on top. The manure heated quickly as it decomposed, warming the soil above enough to stimulate seed germination and seedling growth. A hot bed topped with a cold frame was an excellent way to start tender crops. Today's gardeners have an easy substitute for manure-heated beds: the electric soil-heating cable, available at most garden centers. Spread it in the bottom of your cold frame, plug it in, and the thermostat will automatically maintain a temperature of about 70°F. Arrange your flats of tomatoes, peppers, eggplants, and other tender seedlings on top, and watch them grow. A heating cable keeps a cold frame warm enough for tender seedlings even when the temperature

A combination of mulch and row covers can extend the growing season for carrots and other fall crops in the garden. Or you can grow them in a cold frame.

Extending the Season CONTINUED

Varieties for Short-Season Gardens

In northern and high-mountain gardens, try planting early varieties such as Royal Burgundy snap bean, Red Ace beet, Emerald Acre cabbage, Earlivee sweet corn, Black-seeded Simpson or Red Sails lettuce, Ace bell peppers, and Sweet 100 cherry tomatoes.

1 *To protect individual seedlings from frost, cover them with plastic milk jugs. Weight down the jugs with stones or bricks in windy weather. Remove the jugs during the day.*

2 *When you expect frost, drape floating row covers loosely over plants and weight down the edges with bricks or stones. Roll back the covers the next morning.*

drops well below freezing. On especially chilly nights, toss an old blanket over the cold frame for insulation.

To protect tender plants after transplanting them into the garden, what you need is the equivalent of many small cold frames. Empty plastic milk jugs or 2-liter soft-drink bottles work well. Remove the cap and cut off the bottom; then push the container in place over the plant. Heap soil around the outside to keep the jug from blowing away. Hot caps made from heavy waxed paper work similarly, as do old-fashioned glass bell jars and cloches. Any of these coverings will protect individual plants from a few degrees of frost, but all are liable to overheat on sunny days. Remove them as soon as frost danger is past.

One of the best devices for plant protection is made of two layers of clear plastic that are fused into a series of tubes. When you fill the tubes with water, the protector stands up by itself. You can position it upright, making a cylinder that's open on the top and bottom ends, or lean the top inward to make a tepee. Either way, the water absorbs solar heat during the daytime, then releases it at night, so the temperature inside doesn't fluctuate as extremely as it does inside an old milk jug. This kind of protector is available from mail-order suppliers and seed catalogues, and can be reused for many years. If you can't find one, though, you can borrow the idea by surrounding a seedling with a fort of water-filled jugs.

If protecting individual plants is too much trouble, you can make a tunnel or tent big enough to cover a whole row or bed. Use hoops of wire or plastic pipe for support, putting them about 4 feet apart, then

Four vegetables that taste better after they have been exposed to frost are, clockwise from top left, Brussels sprouts, Russian kale, carrots, and collard greens.

TIMESAVING TIP

Get a head start by pre-sprouting seeds of peas, beans, or corn. Fold the seeds in a damp paper towel, wrap in a plastic bag, and put in a warm, dark place indoors. Check daily, and as soon as the seeds sprout, sow them in the garden.

Extending the Season CONTINUED

How to Predict When You'll Get Frost

Frost is most likely when the sky is clear and the air is still and dry, with no breeze and low humidity. Under these conditions, if the temperature is low already and drops quickly at sunset, cover any tender plants.

cover with row-cover fabric or a sheet of clear plastic (cut slits or holes in the plastic to vent excess heat). Use stones, boards, or soil to hold down the edges.

To make a tent, install two tall stakes with a rope or wire stretched between them like a clothesline. Throw a sheet of clear plastic over the line to form a tent that covers the plants. Peg down the edges, or weight them down with soil or rocks. Open or remove plastic tunnels and tents on mild, sunny days so that the temperature does not rise too high inside and cook your plants.

If you can't resist the gamble of planting tender crops before the average last frost date, and don't want to cover them all with protectors, at least keep a supply of emergency covers on hand. When late frost threatens, you can protect seedlings and transplants with a temporary mulch of hay or straw, or spread old sheets or blankets over them. Don't spread plastic directly over plants—it doesn't offer much protection, and any leaves that touch the plastic will freeze. (Plastic works only as a tent, to trap warm air close to the ground.) Soil itself also offers frost protection. Use a hoe to quickly mound soil over emergent shoots of corn, potatoes, and beans; they'll grow up through it a few days later, unharmed.

Aside from frost protection, another strategy for gaining an early harvest is choosing short-season varieties that germinate and grow quickly even in cool soil. Check the "number of days to maturity" in seed-catalogue descriptions as you choose which varieties to order. There are short-season varieties of snap beans, sweet corn, tomatoes, and several other popular vegetables.

Short-season varieties, and regular ones too, grow fastest if you preheat the soil by covering it with clear or black plastic for a week or two before planting. You can remove the plastic to preheat another area, or punch holes and plant right through it.

The easiest way to extend the season in fall is simply to cover tender plants such as tomatoes and peppers with blankets or other covers for protection from the first cold snap. Often the first one or two frosts are followed by a spell of mild weather, and crops protected once may continue bearing for several weeks before the next, often fatal, cold wave strikes. When you anticipate a serious, killing frost, pick all remaining green or semi-ripe tomatoes, peppers, beans, squash, cucumbers, and other hot-weather crops for safe storage inside.

To gain a much longer extension in fall, plan ahead in summer and plant special fall crops. Work up the soil in your cold frame, and plant a variety of salad greens inside. If you especially like fresh salads, plant extra beds and plan to cover them with tents. Lettuce, spinach, radicchio, endive, arugula, Chinese cabbage, edible chrysanthemum, corn-salad, and French sorrel all reach perfection in cool weather, with much more succulent textures and pleasing flavors than you get in hot weather. However, to pick salads in cool weather, you need to sow the seeds about two months before the first frost. Depending on what you plant, the plants will need at least that much time to reach mature size. They won't grow much more when the days turn short and cold, but they'll keep very well under shelter.

Vegetables Whose Flavor Improves with Frost

Kale, collards, cabbage, lettuce, and other vegetables with edible leaves become milder and lose any bitterness after a few weeks of cool fall weather. Leeks, carrots, beets, and celeriac develop a surprising sweetness. Brussels sprouts also taste sweeter after they've been exposed to frost.

1 *To form a simple plastic tunnel, make a series of metal hoops. Tie strings to stakes at each end of the bed and run longitudinally along the hoops for extra support.*

2 *When frost threatens, stretch a sheet of polyethylene over the hoops. Secure it at both ends, and mound up soil along the sides to seal the tunnel.*

Midsummer plantings of root crops such as carrots, beets, and turnips also provide a delicious harvest throughout fall. Like salad greens, the flavor of these vegetables actually improves during cool weather. You don't know how good carrots are until you taste one that's been pulled in late fall. Its sweetness will amaze you.

If you have a large south window that gets sun all day, you can grow a few treats indoors in winter. A few pots of parsley, dill, chives, and other culinary herbs will add fresh green flavor to all you cook. They won't last through the winter, but they're good for a while. Better yet, start a cherry tomato from seed in midsummer, and raise the seedling in a pot on the patio. Stake it well, and fertilize regularly to keep it vigorous and healthy. Bring it indoors before cold weather, and give it a little shake every day to pollinate the flowers. You won't pick baskets of tomatoes, but each one will be delicious. Bell peppers, chili peppers, and cucumbers can be grown in the same way.

Preparing for Winter

*A*fter hard frost has killed all the tender plants, it's time for fall cleanup in the garden. Discard any diseased plants such as blighted tomato vines. Uproot plants that were healthy, and add them, roots and all, to the compost pile. Chop coarse stems and stalks into short pieces with pruning shears or loppers, or use a power shredder if you have one. Layer these garden wastes with lawn clippings, hay, autumn leaves, farm manure, and other material to make a big compost pile. Monitor the temperature as it heats up, and turn the pile as needed. Cover the finished compost with a tarp to protect it from winter rains and snow.

By the end of the season, most mulches will have sufficiently decomposed for you to till them easily into the soil. Some gardeners prefer to rake up the residue and add it to the compost pile instead. Either way, do something to disturb the mulch and expose insect pests that tried to burrow in for the winter.

Choose one bed for early spring crops, and prepare the soil in advance by digging it well and adding a generous dose of compost. Cover with evergreen boughs to protect the soil from compaction and erosion. (It's easier to remove these boughs in spring than to pull back a layer of heavy, wet straw or leaves.) If you like spinach, plant a small patch in this bed, no later than a few weeks after first frost. The seedlings will overwinter if protected and provide your first crop of spring greens.

Choose other parts of the garden for warm-weather crops, and plant cover crops there to protect the soil in winter and add organic matter. For maximum growth, plant cover crops as soon as space is available, preferably in late summer or early fall. However,

You can leave carrots in the garden all winter and dig them up when you want them. Cover them with a foot of loose mulch. Pull the mulch aside to dig up carrots, then replace it. Mark the location of mulched crops with tall stakes that will stand out in snow.

1 *Leeks, carrots, and some other root vegetables can be left in the garden all winter if well mulched and dug up as needed. Straw or shredded leaves work best as mulch.*

2 *Spread loose mulch a foot or more deep, or stack bales of straw around the plants. Pull aside the mulch and dig up the vegetables when you want them.*

even late plantings will germinate and overwinter as seedlings, then make plenty of new growth in spring before you have to turn them under.

Protect carrots, leeks, parsnips, and salsify for winter harvest by covering them with bags of leaves, bales of hay or straw, or loose straw or hay 1 foot deep. A thick covering will keep the ground from freezing in all but the most severe climates.

Inspect the crops that you plan to store indoors. Set aside bruised, damaged, or not fully mature vegetables for prompt use. Different crops require different storage conditions. Carrots, beets, potatoes, celery, and cabbage need a cool (40°F), dark, damp root cellar. Onions need cool temperatures and darkness, with dry rather than humid air. Sweet potatoes, winter squash, and pumpkins keep best in a semi-heated (50° to 60°F), dark, dry room.

Clean and sort seeds that you are saving to plant next year. Label them before you forget what's what, and store the packets in a cool, dry place.

Gather tomato stakes and cages, trellises, bean poles, fencing, and other pieces of equipment, and store them. Clean soil off shovels, hoes, and other hand tools, and put them inside for winter sharpening and care. Run the gas tank dry and change the oil on power tillers, shredders, and other small-engine tools before putting them into storage.

When the chores are finished, get out your garden notebook and start assessing the performance of this year's garden. It's not too early to begin making notes on what changes you want to make next year, which new vegetables to try, and which seeds and plants to order when the blizzard of catalogues arrives with the winter snow.

Regional Calendar of Care

Spring   *Summer*

COOL CLIMATES

Spring

- About eight weeks before expected last frost, start seedlings of tomatoes, peppers, cabbage, broccoli, and cauliflower under lights indoors.

- As soon as the soil can be worked, sow seeds of peas, spinach, leaf lettuce, carrots, and radishes in the garden.

- Mow cover crops and let the clippings and stubble dry for a few days, then till or fork the garden. Wait at least two weeks for the stubble to start decomposing before planting vegetable crops.

- Harden off seedlings by putting them outdoors in the shade for a few hours a day, then putting them in the sun all day, and finally leaving them out day and night.

- Use clear or black plastic to prewarm the soil before planting heat-loving crops like melons.

Summer

- Harvest peas and other early crops as soon as they ripen, then pull out the plants and sow something else in the same space. Snap beans make a good second crop that matures quickly.

- Start fall crops of broccoli, cauliflower, cabbage, and Brussels sprouts by sowing seeds in a nursery bed about three months before the average date of the first frost.

- Monitor for pests, and treat problems as they occur. Use a fence or traps to restrict raccoons, rabbits, rodents, and deer.

- Water regularly during dry spells. Most crops need an average of about 1 inch of water a week, from rain or irrigation. Don't wait until plants wilt; water as soon as the soil dries out.

- Apply mulch to help retain soil moisture and control weeds. Spread a thick layer over paths, between rows, and around hills.

WARM CLIMATES

Spring

- Get an early start with cool-weather crops. Sow peas, salad greens, carrots, beets, and turnips as soon as daytime highs reach above 60°F. Plant onion sets and potatoes at the same time.

- Prepare the rest of the garden for planting warm-weather crops. It's a good idea to till early, let annual weeds sprout, then hoe or cultivate to eliminate them.

- When danger of frost is past, set out transplants of tomatoes and peppers. Sow beans, squash, cucumbers, okra, Southern peas, melons, and corn.

- Use a few sheets of newspaper or cardboard, topped with bark chips, pine needles, straw, or hay to smother bindweed, Bermuda grass, and other invasive perennial weeds.

Summer

- Begin watering regularly when hot weather comes. Use soaker hoses or a drip system to put water right in the soil, where it's needed, or use flood irrigation to soak between the rows.

- Use plenty of mulch to retain soil moisture and shade plant roots from the heat. Mulch decomposes fast in hot weather. Add more as needed.

- Pick regularly to prolong the harvest from squash, cucumbers, pole beans, and okra. Top-dress these and other crops with compost or fertilizer.

- Plant subsequent crops of beans, squash, tomatoes, and other crops to replace early plantings that peter out.

🍂 Fall

- Use old sheets and blankets, cardboard boxes, row-cover fabric, or whatever else you can find to cover tomatoes, peppers, beans, and other plants for protection from early light frosts.

- Harvest potatoes, sweet potatoes, onions, squash, and pumpkins for winter storage indoors.

- Check weekly for cabbageworms on fall crops of cabbage and related plants. Control with a dusting or spray of Bt. Watch for aphids on spinach and other greens, and spray with soapy water.

- Sow hairy vetch, crimson clover, winter rye, and other cover crops to fill empty spaces after vegetable crops have been harvested and pulled.

- Use bagged leaves or bales of hay or straw to cover rows of carrots and leeks and preserve them for winter harvest.

🌲 Winter

- After a hard frost, clean up all remaining crop debris, any weeds, and other plant litter. Chop coarse material for faster composting.

- Make and tend a big hot-compost pile, using garden waste, fall leaves, spoiled hay, farm manure, and whatever else is available. Cover the finished compost with a plastic tarp to protect it from heavy winter rains and snow.

- Till or fork over one raised bed. It will be the first soil to warm up and dry out for early planting in spring.

- Review your garden notebook and make plans for next year's garden. Enjoy studying the new seed catalogues as they arrive, and place your order early for fast delivery.

This table offers a basic outline of garden care by season. The tasks for each season differ for warm and cool climates: warm climates correspond to USDA Plant Hardiness zones 8 through 11, and cool climates to zones 2 through 7. Obviously, there are substantial climate differences within these broad regions. To understand the specific growing conditions in your garden, consult the Zone Map on page 127. Also be sure to study local factors affecting the microclimate of your garden, such as elevation and proximity of water.

- Add more compost or fertilizer to supply new nutrients when preparing beds for fall planting.

- When temperatures begin to cool, sow carrots, beets, turnips, and other root crops for fall and winter harvest. Sow a variety of salad greens, too.

- Set out transplants of cabbage and other cole crops. Be sure to shade them from the hot sun for a week or so, to keep them from wilting.

- Continue watering regularly.

- Gather cornstalks, bean vines, and other garden waste to make a big compost pile. Chop coarse material and layer with lawn clippings, livestock manure, and fall leaves.

- Sow vetch, clover, or other cover crops to add organic matter to the soil.

- Use row-cover fabrics or plastic tents to protect salad greens from hard frosts, and you can continue picking well into the winter.

- Root crops should hold through the winter without any problem if you keep the soil moist and covered with a medium layer of mulch.

- Review your garden notebook and make plans for next year's garden. Try to plan a strategy of succession planting that will provide for continuous harvest.

- Order seeds early for faster service. Organize your seed-starting pots and trays, and get a supply of soil mix so that you can start sowing indoors soon after New Year's.

Vegetables for American Gardens

*T*his section provides concise information on more than 110 vegetables recommended for American gardens. Favorites like tomatoes and zucchini are covered, as well as less common vegetables such as cardoon and salsify. The table contains descriptions of the plants and the edible parts, growing conditions, directions for planting, and harvesting tips. Each entry includes a photograph of the vegetable.

▼ About Plant Names

Plants appear in alphabetical order by the common name, shown in bold type. On the next line are other widely used common names. Where there is more than one entry for a type of vegetable, the entries are alphabetized by type. For example, you will find lima beans under B rather than L. The third line contains the complete botanical name: genus, species, and where applicable, a variety or cultivar name.

When several species in a genus are similar in appearance and cultural needs, such as garlic and elephant garlic, they are listed together in a single entry in the chart. Most entries, however, describe one particular plant or type of plant.

The second column of the table provides a brief description of the plant and the edible portion. Look here for information on how and when to harvest.

▼ Plant Height

This column of the table lists the average mature height of the plant.

▼ Time to Harvest

Under Time to Harvest you will find the approximate length of time it takes after planting until harvesting can begin. For most vegetables the time to harvest is given in days. Harvest times will vary somewhat according to weather, growing conditions, and the variety being grown.

▼ Planting Guide

The Planting Guide column gives information on how deep to plant, how far apart to space plants within rows or beds, and spacing distance between the rows.

▼ Hardiness

Plants are listed as hardy, half-hardy, or tender. Hardy plants can tolerate quite a bit of frost and can be planted outdoors in early spring as soon as the soil can be worked. Half-hardy plants can tolerate some light frost, but are damaged by prolonged exposure to cold. Plant them out when the danger of heavy frost is past, but occasional light frosts may still occur.

Tender vegetables cannot withstand any frost at all. Do not plant them in the garden until all danger of frost is past (a week or two after the average last frost date for your area) and the soil is warming.

▼ Growing Conditions

The last column of the chart summarizes the best growing conditions for the plant. Look here for information on sun, soil, and moisture needs.

			Plant Height	Time to Harvest	Planting Guide	Hardiness	Growing Conditions
	AMARANTH PURPLE AMARANTH *Amaranthus cruentus* TAMPALA *A. tricolor*	An annual grown for its leaves, which are used in salads or cooked as greens. Cultivars bear long-stemmed, oval, 3- to 6-in. leaves in red or green. Mature plants bear seed-heads, which may be harvested for edible grain.	1½–2½'	30–50 days	Depth: ⅛–¼" Spacing: 4–6" Row spacing: 8–18"	Tender	Full sun. Moist, well-drained soil rich in humus. This rapidly growing plant likes hot, humid weather. Start seeds indoors 4 weeks before last frost or sow outdoors after danger of frost has passed. Pinch off main shoot to increase branching.
	ARTICHOKE GLOBE ARTICHOKE *Cynara scolymus*	A perennial whose large (3- to 5-in.), scaly flower buds are harvested before they are fully developed. At maturity the flowers of this stately plant are a brilliant sky blue, but inedible.	3–6'	150–180 days	Depth: root buds at soil surface Spacing: 3–6' Row spacing: 6–8'	Half-hardy	Full sun. Well-drained, sandy-loam soil that never completely dries out. Plants need a long season to grow to harvest size. In zone 7 and colder dig up roots, remove tops, and overwinter in a cool, dry spot. Propagate from root divisions.
	ARUGULA ROCKET, ROQUETTE *Eruca vesicaria* subsp. *sativa*	An annual member of the mustard family whose leaves have a pungent, nutty flavor. The deeply lobed, 6-in. leaves form mounds of foliage. Harvest leaves and flowers just above soil surface to promote growth; use both in salads.	1–2'	45–60 days	Depth: ⅛–¼" Spacing: 6" Row spacing: 8–12"	Hardy	Full sun to light shade. Moist, well-drained soil. Plants grow best in cool weather. Sow seeds outdoors in early spring for late spring harvest and again in late summer for autumn harvest.
	ASPARAGUS GARDEN ASPARAGUS *Asparagus officinalis*	A perennial whose young shoots (spears) are harvested as they emerge from the soil in spring. When fully developed, the branched shoots have feathery, fernlike, soft leaves. Cream-colored flowers produce red berries.	3–6'	3rd year, then annually in spring	Depth: root crowns 6–8" Spacing: 1½–2' Row spacing: 4'	Hardy	Full sun. Well-drained soil that never completely dries out. Root crowns are male or female; male plants bear more spears. Plant crowns 6–8 in. deep in trenches. Do not harvest until third year. Provide mulch in regions with cold winters.
	ASPARAGUS BEAN YARD-LONG BEAN *Vigna unguiculata* subsp. *sesquipedalis*	A long, twining annual vine. The beans sometimes grow to over 2 ft. long, but fresh pods should be harvested when they are less than 1 ft. long. For dry shell beans, pick pods as they start to turn from green to yellow.	6–8'	50–100 days	Depth: ½–1" Spacing: 4" Row spacing: 4–5'	Tender	Full sun. Well-drained, evenly moist soil, rich in humus. Plants grow best where summer is warm and humid. Sow seeds after soil has warmed and frost danger has passed. Provide supports for vines to grow upon.

Vegetables for American Gardens

			Plant Height	Time to Harvest	Planting Guide	Hardiness	Growing Conditions
BEAN BROAD BEAN, FAVA BEAN *Vicia faba*		A tall European favorite with bushy, thick stems that bear abundant 6- to 8-in. pods containing 1-in. beans. The beans are shelled and used fresh or dried. Pods can be eaten if harvested when half-grown.	3–5'	75–90 days	Depth: 1–1½" Spacing: 4–6" Row spacing: 1½–3'	Half-hardy	Full sun. Average, well-drained soil enriched with compost and kept evenly moist. Plants grow best in cool weather. Sow seeds outdoors in early spring or, in zone 9 and warmer, in winter. Provide supports for plants.
BEAN BUSH LIMA BEAN *Phaseolus limensis* var. *limenanus*		A bush variety of the annual vine. Plants bear large shell beans, which may be used fresh or dried. Pick the flat pods as they start to fill out and bulge. Harvest carefully since the stems break easily.	10–18"	60–85 days	Depth: 1–1½" Spacing: sow 2–3" apart; thin to 6" Row spacing: 2–3'	Tender	Full sun. Well-drained, sandy-loam soil. Do not add additional nitrogen fertilizer. Sow seeds after last frost when the soil is warm; cold soil delays seed germination. Seeds may benefit from dusting with Rhizobium inoculant.
BEAN BUSH SNAP BEAN *Phaseolus vulgaris* var. *humilis*		A variety of bean with edible pods that grows as a low bush rather than a trailing vine. Many cultivars and hybrids provide a choice of pod color (green, wax yellow, or purple), texture, and maturation time.	10–24"	50–70 days	Depth: 1" Spacing: sow 2" apart; thin to 4" Row spacing: 2–3'	Tender	Full sun. Well-drained, slightly acidic, sandy-loam soil. Do not add additional nitrogen fertilizer. Sow seeds after last frost when the soil is warm; cold soil delays seed germination. Seeds may benefit from dusting with Rhizobium inoculant.
BEAN DRY BEAN, SHELL BEAN *Phaseolus vulgaris*		A species that includes many varieties of beans, including those that are shelled and dried, such as the red kidney bean, the large, white 'Great Northern' bean, the speckled tan pinto bean, and the small, white navy bean. Bush cultivars are available.	10–24"	65–100 days	Depth: 1" Spacing: sow 2" apart; thin to 4" Row spacing: 2–3'	Tender	Full sun. Well-drained, slightly acidic, sandy-loam soil. Do not add additional nitrogen fertilizer. Sow seeds after last frost when the soil is warm; cold soil delays seed germination. Seeds may benefit from dusting with Rhizobium inoculant.
BEAN HORTICULTURAL BEAN, SHELL BEAN *Phaseolus vulgaris*		A shell bean cultivar bearing 6- to 8-in. yellow or light green pods, usually streaked with maroon or violet, adding beauty to the garden. The solid-colored or speckled seeds are used fresh or dried. Immature pods can be eaten as snap beans.	1–6'	55–70 days	Depth: 1" Spacing: 2–6" Row spacing: 1–2½'	Tender	Full sun. Well-drained, slightly acidic, sandy-loam soil. Do not add additional nitrogen fertilizer. Sow seeds after last frost when the soil is warm; cold soil delays seed germination. Seeds may benefit from dusting with Rhizobium inoculant.

			Plant Height	Time to Harvest	Planting Guide	Hardiness	Growing Conditions
	BEAN POLE LIMA BEAN *Phaseolus limensis*	*The traditional species of lima bean, grown on long poles or other supports. The pole lima beans need more space in the garden and take longer to mature than bush cultivars, but yield a larger crop.*	8–20'	*85–130 days*	Depth: 1–1½" Spacing: 10–18" Row spacing: 3–4'	Tender	*Full sun. Well-drained, sandy-loam soil. Do not add additional nitrogen fertilizer. Sow seeds after last frost when the soil is warm. Provide supports when seedlings appear. Seeds may benefit from dusting with Rhizobium inoculant.*
	BEAN POLE SNAP BEAN *Phaseolus vulgaris*	*The traditional, vining variety of edible-pod bean requiring poles or fences for support. Many cultivars and hybrids provide a choice of pod color, texture, and maturation time. Pole beans bear later but larger yields than bush beans.*	6–10'	*55–75 days*	Depth: 1–1½' Spacing: 6–8" Row spacing: 3–4'	Tender	*Full sun. Well-drained, sandy-loam soil. Do not add additional nitrogen fertilizer. Sow seeds after last frost when the soil is warm. Provide supports when seedlings appear. Seeds may benefit from dusting with Rhizobium inoculant.*
	BEAN SCARLET RUNNER BEAN *Phaseolus coccineus*	*A perennial vine grown as an annual and bearing brilliant scarlet, showy, 1-in. pealike flowers. The edible young pods and seeds are a bonus. The climbing, twining vines have leaves with three leaflets.*	8–15'	*70–80 days*	Depth: 1–2" Spacing: 6–8" Row spacing: 3–4'	Half-hardy	*Full sun. Moist, well-drained soil rich in humus. Plant seeds in mid-spring a couple of weeks before the last frost. Provide tall supports for the vines to grow on.*
	BEET RED BEET *Beta vulgaris* Crassa group	*A biennial grown for both its bulbous taproot and its edible leaves. Most beet roots are shades of red, but gold cultivars are available. Swiss chard is a member of this species and is grown exclusively for its leaves.*	6–12"	*45–65 days*	Depth: ¼" Spacing: sow 2" apart; thin to 4–6" Row spacing: 2–3'	Hardy	*Full sun. Well-drained, sandy-loam soil. Plants may rot in soggy soil. If soil is clayey, add compost and gypsum. Too much nitrogen fertilizer causes leaves to grow at the expense of the roots. Sow seeds outdoors in early spring.*
	BOK CHOY CHINESE MUSTARD, PAK-CHOI *Brassica rapa* Chinensis group	*An Asian leafy vegetable whose dark green, erect leaves form a loose cluster. Leaves typically have succulent, edible, white stems that overlap at their bases. Harvest by pulling individual leaves or by cutting the entire top at ground level.*	8–20"	*45–60 days*	Depth: ¼" Spacing: sow 1" apart; thin to 6" Row spacing: 1½'	Hardy	*Full sun. Well-drained, evenly moist soil with added compost and lime. Plants grow best in cool weather. For a spring crop sow outdoors in early spring. For an autumn crop sow outdoors in late summer.*

Vegetables for American Gardens

			Plant Height	Time to Harvest	Planting Guide	Hardiness	Growing Conditions
	**BROCCOLI** *Brassica oleracea* Botrytis group	A biennial grown as an annual. Sturdy stems produce rounded heads of dark green flower buds that are eaten cooked or raw. Harvest heads before the yellow flower petals begin to show. Some cultivars are purple or chartreuse.	1½–3'	70–95 days	Depth: ¼–½" Spacing: 1½–2' Row spacing: 1½–3'	Hardy	Full sun. Well-drained, evenly moist soil with added compost and lime. Plants grow best as a cool-weather spring crop. Start seeds indoors 4–6 weeks before last frost. For an autumn crop sow seeds outdoors 10 weeks before first frost.
	**BROCCOLI RABE** TURNIP BROCCOLI *Brassica rapa* Ruvo group	A close relative of the turnip, grown not for its tap-root, but for the glossy leaves and sharp, sweet flower heads, which are cooked as greens or used raw in salads. This is a favorite in Italian gardens.	12–18"	45–65 days	Depth: ¼–½" Spacing: 4–6" Row spacing: 1½–2'	Hardy	Full sun. Well-drained, evenly moist soil with added compost and lime. Plants grow best as a cool-weather spring crop. Start seeds indoors 4–6 weeks before last frost. For an autumn crop sow seeds outdoors 10 weeks before first frost.
	**BRUSSELS SPROUTS** *Brassica oleracea* Gemmifera group	A biennial grown as an annual. The flowering branches form buds (sprouts) that look like miniature cabbages. Harvest sprouts by snapping them off. This is one of the most cold hardy of the "cole" or cabbage-related crops.	2–4'	90–120 days	Depth: ¼–½" Spacing: 1½–2' Row spacing: 2–3½'	Very hardy	Full sun. Evenly moist, well-drained soil with added compost and lime. Start seeds indoors in spring, 4–6 weeks before last frost. Harvest sprouts in late summer to fall as needed from bottom of stem toward top. They taste sweeter after frost.
	**CABBAGE** *Brassica oleracea* Capitata group	An annual bearing a succulent-leaved head that conceals the flowering parts of the plant. Harvest early types before they "bolt" and elongate in summer. Harvest late types before the hard freezes of autumn; these are best for winter storage.	12–15"	65–125 days	Depth: ¼–½" Spacing: 10–24" Row spacing: 2–3½'	Hardy	Full sun. Well-drained, evenly moist soil with added compost and lime. Grows best in cool weather. For a spring harvest start seeds indoors 4–6 weeks before last frost. For an autumn crop sow outdoors 10 weeks before first frost.
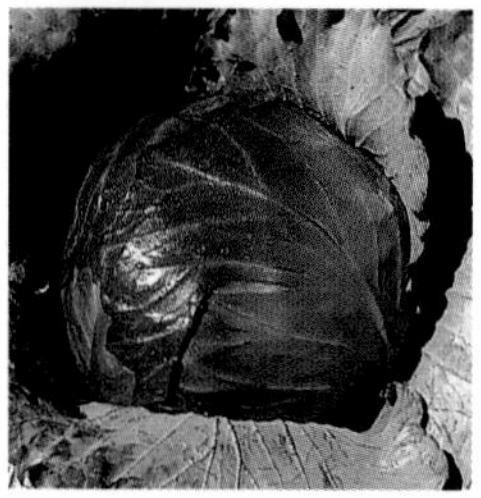	**CABBAGE** RED CABBAGE *Brassica oleracea* Capitata group	Red or purple cultivars of green cabbage. These are used raw in salads or as a cooked vegetable. Early and late cultivars are available.	12–15"	65–125 days	Depth: ¼–½" Spacing: 10–24" Row spacing: 2–3½'	Hardy	Full sun. Well-drained, evenly moist soil with added compost and lime. Grows best in cool weather. For a spring harvest start seeds indoors 4–6 weeks before last frost. For an autumn crop sow outdoors 10 weeks before first frost.

			Plant Height	Time to Harvest	Planting Guide	Hardiness	Growing Conditions
	CABBAGE SAVOY CABBAGE *Brassica oleracea* Capitata group	*A cabbage form whose heads have thin, crinkled leaves with a waffle-like texture. The attractive, tender leaves are often dark green with light green veins. As with other cabbages, savoy is used raw or cooked.*	12–15"	*65–125 days*	Depth: 1/4–1/2" Spacing: 10–24" Row spacing: 2–3 1/2'	Hardy	*Full sun. Well-drained, evenly moist soil with added compost and lime. Grows best in cool weather. For a spring harvest start seeds indoors 4–6 weeks before last frost. For an autumn crop sow outdoors 10 weeks before first frost.*
	CARDOON *Cynara cardunculus*	*A close relative of the globe artichoke and a perennial that is grown for its edible leaf bases and stalks. Like its relative, cardoon has gray-green, clasping leaves and cobalt blue flowers. To harvest stalks cut the stem below the ground.*	6–8'	*150–180 days*	Depth: 1/2" Spacing: 18–24" Row spacing: 4–5'	Half-hardy	*Full sun. Well-drained, sandy-loam soil that never completely dries out. Plants need a long season to mature. Propagate from root divisions or seeds. Grow in trenches; fill in trench as plant grows to blanch stem bases.*
	CARROT *Daucus carota* var. *sativus*	*A biennial grown for its taproot. Many cultivars are available with varying lengths and shapes. Harvest during the first year after the root has developed but before winter.*	8–12"	*55–75 days*	Depth: 1/4–1/2" Spacing: sow 1" apart; thin to 3–4" Row spacing: 12–15"	Hardy	*Full sun. Well-drained, rock-free, sandy-loam soil that is slightly acidic. Carrots need excellent drainage and tend to rot in soggy conditions. In early spring, sow outdoors in raised beds to give warmth to seeds that are slow to germinate in cold soil.*
	CAULIFLOWER *Brassica oleracea* Botrytis group	*A cold-season biennial grown as an annual. It forms an enormous, curdlike cluster of flower buds. Usually the heads are white, but purple cultivars are available. Harvest heads before they bolt and start to flower.*	18–24"	*50–90 days*	Depth: 1/4–1/2" Spacing: 1–1 1/2' Row spacing: 2–3'	Hardy	*Full sun. Moist, well-drained alkaline soil; add lime if needed. Plants grow best in cool weather. For a spring crop sow seeds indoors 6–8 weeks before last frost. For an autumn harvest sow seeds outdoors 8–10 weeks before first frost.*
	CELERIAC *Apium graveolens* var. *rapaceum*	*A biennial grown as an annual for its white, bulbous, 3- to 5-in. root that tastes like mild celery. This variety of celery also has ribbed stems and leafy stalks.*	10–14"	*90–120 days*	Depth: 1/8" Spacing: 6–12" Row spacing: 2–3'	Half-hardy	*Full sun. Humus-rich, moist, well-drained soil. Grow in raised beds to encourage good drainage. Start seeds indoors 10–12 weeks before last frost; transplant in early spring. Grows best in warm, but not excessively hot, weather.*

Vegetables for American Gardens

			Plant Height	Time to Harvest	Planting Guide	Hardiness	Growing Conditions
	CELERY Apium graveolens var. dulce	A familiar biennial grown as an annual for its ribbed, overlapping, leafy stalks. Some cultivars are self-blanching, while others need soil mounded around their stems as they grow.	15–18"	80–110 days	Depth: 1/8" Spacing: 6–8" Row spacing: 2–3'	Half-hardy	Full sun. Humus-rich, moist, well-drained soil. Grow in raised beds to encourage good drainage. Start seeds indoors 10–12 weeks before last frost; transplant in early spring. Grows best in warm, but not excessively hot, weather.
	CHARD SWISS CHARD Beta vulgaris Cicla group	A biennial grown as an annual for its red- or white-stemmed leaves with crin-kled edges. Swiss chard is related to beets. Its leaves are cooked as greens or used fresh in salads. Snap off the leaves at the base; new ones will sprout.	1–2'	50–65 days	Depth: 1/3–1/2" Spacing: sow 1" apart; thin to 6" Row spacing: 1 1/2–2'	Hardy	Full sun to light shade. Well-drained, sandy-loam soil. If soil is clayey, add compost and gypsum. Sow seeds out-doors where desired from early to late spring.
	CHICK PEA GARBANZO BEAN Cicer arietinum	A bushy annual legume bearing small leaves with 9–15 leaflets and 1-in., swollen seedpods. Each pod contains several round, starchy seeds with a nutty flavor. Harvest seeds when pods are dry and start to split open.	1–2'	85–90 days	Depth: 1" Spacing: 8–12" Row spacing: 2–2 1/2'	Tender	Full sun. Well-drained, sandy-loam soil. Sow seeds outdoors after all danger of frost has passed. Plants grow best in hot, dry weather.
	CHICORY BELGIAN ENDIVE, FORCING CHICORY, WITLOOF Cichorium intybus	A perennial grown in the garden to produce roots that are then brought indoors to be forced. The resulting crisp, white-green, 4- to 6-in. leaves are used as a mild-ly bitter green in winter sal-ads or in appetizers.	6–8"	110–300 days	Depth: 1/4–1/2" Spacing: 8–12" Row spacing: 1–1 1/2'	Hardy	In autumn cut tops of mature plants back to 1 in., lift plants, and trim roots to 8 in. Store horizontally in sand in a cool, dark spot. To force, plant roots vertically in sand with tops 6 in. below surface. Keep moist, dark, and cool for 2–3 weeks.
	CHICORY CUTTING CHICORY, SUGAR LOAF CHICORY Cichorium intybus	A perennial that produces upright, clustered heads that resemble romaine lettuce. It tastes slightly bitter, like its close relatives Belgian endive and radicchio. Harvest before it produces its blue, dandelion-like flowers.	10–18"	70–110 days	Depth: 1/4–1/2" Spacing: 6–12" Row spacing: 1–2'	Hardy	Full sun to light shade. Well-drained, fertile soil rich in humus. Keep the soil evenly moist to prevent leaf wilting. Sow seeds outdoors in mid-spring, even before the last frost.

		Plant Height	Time to Harvest	Planting Guide	Hardiness	Growing Conditions
CHINESE CABBAGE *Brassica rapa* Pekinensis group	A cabbage available in heading and loose-leaved forms. The wrinkled, crisp leaves typically have white veins and a mild flavor. Harvest by cutting the stem 1 in. above the soil; often the plants will resprout for a second crop.	12–18"	65–80 days	Depth: ¼" Spacing: 9–12" Row spacing: 1½–2'	Half-hardy	Full sun. Fertile, well-drained, evenly moist soil. This annual grows best in cool weather; it sometimes bolts in hot weather. For a spring crop sow seeds indoors in winter. For an autumn crop sow outdoors in mid- to late summer.
CHINESE KALE CHINESE BROCCOLI *Brassica oleracea* Alboglabra group	A favored vegetable in Asian cooking. This annual has succulent, blue-green leaves, stems, and heads. Harvest before the heads "bolt" and reveal the yellow, 4-petaled flowers.	12–18"	65–70 days	Depth: ¼" Spacing: 9–12" Row spacing: 1–2'	Half-hardy	Full sun. Fertile, well-drained, evenly moist soil. Sow seeds outdoors in late spring to late summer. Harvest main stems first and then the lateral stems, which enlarge later.
CHRYSANTHE-MUM GARLAND CHRYSAN-THEMUM, SHUNGIKU *Chrysanthemum coronarium*	An edible, yellow- or orange- flowered annual chrysanthemum used widely in Asian cuisine. Its deeply cut, aromatic, gray-green leaves usually are harvested when plants are 4 to 8 in. high; they are used as greens in stir-fry dishes.	2–4'	40–50 days	Depth: ¼" Spacing: sow 1" apart; thin to 3" Row spacing: 1½'	Very hardy	Full sun to light shade. Moist, well-drained soil that is rich in humus. Plants grow best in cool weather. Sow seeds outdoors in very early spring and then successively into early summer.
COLLARDS COLLARD GREENS *Brassica oleracea* Acephala group	A biennial related to cabbage and grown for its loose open rosettes of succulent leaves. Harvest the gray-green, spatula-shaped leaves when they are small. Do not take the "heart" leaves at the center of the clump until new ones are formed.	2–4'	60–90 days	Depth: ¼–½" Spacing: sow 1" apart; thin to 9" Row spacing: 1½–3'	Very hardy	Full sun to light shade. Well-drained, evenly moist soil with added compost and lime. Plants grow best in cool weather. For a spring crop sow seeds outdoors in early spring. For an autumn crop sow outdoors in late summer.
CORN POPCORN *Zea mays* var. *praecox*	A native American crop that takes up a lot of space in the garden, but is fun to grow and more fun to eat. Harvest when ears are hard and glossy (may mature after frost). Cure ears in a dry place for several weeks after cutting them from stalk.	4–7'	85–115 days	Depth: 1" Spacing: sow 3" apart; thin to 9–12" Row spacing: 2–3'	Tender	Full sun. Well-drained soil that is evenly moist. Plant in blocks of 4 rows to ensure pollination. Allow at least 25 ft. between different varieties of corn to prevent cross-pollination. Corn grows best in warm weather.

Vegetables for American Gardens

			Plant Height	Time to Harvest	Planting Guide	Hardiness	Growing Conditions
	CORN SWEET CORN *Zea mays var. rugosa*	A tall American native grass. Many feel that the taste of sweet corn directly from the garden more than justifies the space it requires. Cultivars are available in a range of sweetness. Harvest when juice from kernels is milky.	4–7'	85–115 days	Depth: 1" Spacing: sow 3" apart; thin to 9–12" Row spacing: 2–3'	Tender	Full sun. Well-drained, evenly moist soil. Plant in blocks of 4 rows to ensure pollination. Allow at least 25 ft. between different varieties. Successive plantings from mid-spring to early summer will prolong harvest.
	CORN-SALAD LAMB'S-LETTUCE, MACHE *Valerianella locusta*	A hardy annual grown for its rosettes of spatula-shaped, mild-tasting, dark green leaves. Plants emerge in early spring and are used for salad greens. Harvest leaves before they reach 4 in.	3–6"	45–60 days	Depth: 1/4" Spacing: 1–2" Row spacing: 1–1 1/2'	Hardy	Full sun to light shade. Moist, well-drained soil. Plants grow best in cool weather and will overwinter even in areas with cold winters. Sow outdoors in early spring and late summer.
	CUCUMBER PICKLING CUCUMBER *Cucumis sativus*	A vining annual that produces small yellow flowers and blocky, 2- to 6-in. fruits. Harvest when cucumbers are small and tender. Harvest frequently; a single mature fruit left on the vine may cause the entire plant to cease blooming.	4–6'	50–75 days	Depth: 1/2" Spacing: sow 3" apart; thin to 6–9" Row spacing: 5–6'	Tender	Full sun. Well-drained soil rich in humus and lime. Plants grow best in warm weather. Sow seeds after soil is warm, several weeks after last frost. Provide supports for emerging vines.
	CUCUMBER SLICING CUCUMBER *Cucumis sativus*	A classic salad vegetable with cylindrical, 6- to 10-in., deep green fruits. To ensure continued fruiting, harvest maturing cucumbers daily before their skins start to yellow. This annual vine can be trained to grow on a trellis.	4–6'	50–70 days	Depth: 1/2" Spacing: sow 3" apart; thin to 6–9" Row spacing: 5–6'	Tender	Full sun. Well-drained soil rich in humus and lime. Plants grow best in warm weather. Sow seeds after soil is warm, several weeks after last frost. Provide supports for emerging vines.
	DANDELION *Taraxacum officinale*	A familiar perennial wildflower available in cultivars that produce better tasting greens than wild types. Mature leaves are bitter. Young leaves are mild and can be used in salads or as cooked greens if harvested in early spring.	3–9"	90–100 days	Depth: 1/4–1/2" Spacing: 12–14" Row spacing: 1–1 1/2'	Very hardy	Full sun. Rich, moist soil. The first year start indoors in late winter and transplant in spring. Harvest leaves in spring and again in autumn. Remove flower heads to boost leaf production. Plants can be forced indoors for greens in winter.

			Plant Height	Time to Harvest	Planting Guide	Hardiness	Growing Conditions
	EGGPLANT AUBERGINE *Solanum melongena* var. *esculentum*	A perennial grown as an annual for its 4- to 10-in. glossy, deep purple, green, or white fruits, which are eaten cooked. All other parts of plant are toxic. Harvest before fruits reach full size by pruning stem with a knife or pruning clippers.	1½–2'	60–85 days	Depth: ¼–½" Spacing: 1½–2' Row spacing: 2–4'	Tender	Full sun. Fertile, well-drained soil that is kept evenly moist. Start seeds indoors 8–10 weeks before last frost. Soak seeds overnight to encourage germination. Transplant when nighttime temperatures are above 50°F.
	ENDIVE CURLY ENDIVE, ESCAROLE *Cichorium endivia*	A slightly bitter salad green used extensively in European cuisine. Endive is the curly-leaved form and escarole the broad, flat-leaved form. Some cultivars are self-blanching. Escarole has a milder flavor. Light frost improves the flavor of both.	8–14"	45–100 days	Depth: ¼–½" Spacing: 8–15" Row spacing: 1–1½'	Hardy	Full sun to light shade. Fertile, well-drained soil that is evenly moist. To grow as a fall crop sow seeds outdoors in midsummer in zone 6 and cooler. For a late winter crop in zone 7 and south, sow outdoors in late autumn.
	FENNEL FLORENCE FENNEL, FINOCCHIO *Foeniculum vulgare* var. *azoricum*	A variety of herb grown for its swollen, clasping leaf bases that form a "bulb," eaten raw or cooked. Both finocchio and the related herb, culinary fennel, bear many feathery leaves with an anise flavor. Harvest whole top; leaves resprout.	1½–2'	90–100 days	Depth: ¼–½" Spacing: 5–7" Row spacing: 1–1½'	Half-hardy	Full sun. Sandy, well-drained, evenly moist soil that is fertile and slightly alkaline. Plants grow most vigorously in warm weather. Sow seeds outdoors in mid-spring where plants are desired.
	GARDEN CRESS PEPPERGRASS *Lepidium sativum*	An annual herb grown for sprouts and for its peppery tasting leaves, which cluster around the base of the stem. Sprouts and leaves are used as a salad seasoning. Cultivars with curly, parsley-like leaves are available.	1–2'	10–12 days for sprouts; 45–60 days for greens	Depth: ⅛–¼" Spacing: 1–3" Row spacing: 1'	Hardy	Full sun. Ordinary garden soil. Grows best in cool weather. Sow new seeds outdoors or as pot plants at monthly intervals starting in early spring and harvest before flowering starts.
	GARLIC *Allium sativum* ELEPHANT GARLIC *A. scorodoprasum*	Pungent bulb-forming perennials. The leaves can be used as scallions. The bulbs consist of many sheathed scales or "cloves." The larger elephant garlic has milder cloves that are often covered with purple skin.	1–3'	90–100 days	Depth: 1–2" Spacing: 4–6" Row spacing: 12–15"	Hardy	Full sun. Fertile, well-drained, evenly moist soil. For best results grow as a perennial. Plant "sets" (small bulbs) several weeks before first frost. The next year pinch flowers to promote bulbs and harvest in autumn when tops die back.

Vegetables for American Gardens

			Plant Height	Time to Harvest	Planting Guide	Hardiness	Growing Conditions
GOOD-KING-HENRY *Chenopodium bonus-henricus*		A European potherb grown for its young shoots and spinachlike leaves. Sometimes this perennial is grown in beds, and its emerging shoots are harvested like asparagus in the spring.	2–3'	70–90 days	Depth: 1/8" Spacing: 6–12" Row spacing: 1 1/2–2'	Hardy	Full sun. Rich, moist soil with added compost. Sow seeds outdoors in early spring by pressing them gently into the soil surface. Harvest by thinning and by cutting branch tips, which stimulates resprouting.
HORSERADISH *Armoracia rusticana*		A perennial grown mainly for its white, fleshy, sharply pungent roots. Mashed and used as a condiment, a little of this perennial goes a long way. Young leaves can be used in salads; older leaves are large and tough.	2–2 1/2'	120–150 days	Depth: 2–4" Spacing: 1–1 1/2' Row spacing: 3–4'	Hardy	Full sun. Well-drained soil. Propagate from root cuttings only. Dig plants and harvest roots in autumn. Plants may become invasive. To check growth, remove all roots in autumn; replant a few roots in spring.
JERUSALEM ARTICHOKE SUNCHOKE *Helianthus tuberosus*		A perennial species of sunflower. Plants produce tubers, which have the texture of potato but a flavor all their own. They are used raw in salads or cooked. Tall stems produce 3- to 4-in. sunflowers.	6–8'	120–150 days	Depth: 3–4" Spacing: 1–1 1/2' Row spacing: 3'	Hardy	Full sun to light shade. Well-drained, fertile, humus-rich soil. Plant fresh pieces of tuber, each with at least 1 bud, in the early spring. Harvest tubers in autumn. This perennial can become weedy.
KALE *Brassica oleracea* Acephala group		A nonheading relative of cabbage that is cooked as greens or used as a garnish. Kale produces thick leaves on an erect stem. Harvest leaves from the middle to the top of the stem. Flavor of leaves is best after frost.	1–1 1/2'	60–70 days	Depth: 1/2" Spacing: sow 6" apart, thin to 2' Row spacing: 2–4'	Hardy	Full sun to light shade. Well-drained, evenly moist soil with added compost and lime. Sow directly outdoors in early spring. Plants grow best in cool weather; harvest leaves in late spring and again in autumn.
KALE FLOWERING CABBAGE, FLOWERING KALE *Brassica oleracea* Acephala group		Ornamental, edible relative of the familiar garden cabbage. Plants produce rosettes of highly decorative, colorful leaves rather than tight heads. Cultivars range from white to green to red and purple in various leaf texture.	8–12"	60–70 days	Depth: 1/2" Spacing: 8–12" Row spacing: 1 1/2–2'	Very hardy	Full sun. Moist, well-drained soil. Plants grow best in cool weather; colors improve with frost. For best color, do not fertilize. Sow seeds outdoors in mid- to late summer. Protect from cabbage butterfly caterpillars.

			Plant Height	Time to Harvest	Planting Guide	Hardiness	Growing Conditions
	**KOHLRABI** *Brassica oleracea* Gongylodes group	A plant whose turniplike, bulbous stems taste like its close relative, cabbage. "Bulbs" are used raw or cooked and are sweetest when they are 2 in. or less in diameter. Purple- or white-stemmed cultivars are available.	1–1½'	40–60 days	Depth: ¼–½" Spacing: 4–5" Row spacing: 1–2'	Half-hardy	Full sun. Moist, fertile, well-drained soil. Plants grow best in cool weather, but withstand heat if soil is sufficiently moist. Sow seeds outdoors at 2-week intervals starting in early spring. Harvest whole plant.
	**LEEK** *Allium ampeloprasum* Porrum group	A perennial onion with an exquisitely mild taste. Inch-wide, flattened, clasping leaves overlap a cylindrical flower stalk; leeks do not form bulbs. Mound soil over the growing stalks to blanch their bases.	2–3'	60–110 days	Depth: ¼–½" Spacing: 4–6" Row spacing: 2–2½'	Half-hardy	Full sun. Well-drained, evenly moist soil that is rich in humus. Start seeds indoors 10–12 weeks before last frost, or sow seeds outdoors in early spring. Leeks are slow growers. Harvest in autumn after frost.
	**LETTUCE** BUTTERHEAD LETTUCE *Lactuca sativa*	A cool-season annual whose leaves are used as salad greens. The large, tender leaves are arranged in open rosettes or loose heads. This group includes the familiar 'Boston', 'Bibb', and 'Buttercrunch' lettuces.	6–12"	50–80 days	Depth: ¼–½" Spacing: 6–9" Row spacing: 1–1½'	Half-hardy	Full sun to very light shade with some afternoon shade in warm climates. Well-drained soil that never completely dries out. Sow seeds outdoors from early spring to late summer at 2-week intervals. Harvest before leaves become bitter.
	**LETTUCE** COS LETTUCE, ROMAINE LETTUCE *Lactuca sativa*	A slow-growing gourmet lettuce with upright, oval heads and crisp, thick, sweet leaves that are quite tender. Cultivars have colors ranging from dark green to light green to red. Thin young seedlings and use as leaf lettuce.	8–12"	45–50 days for leaves; 60–90 days for heads	Depth: ¼–½" Spacing: sow 1" apart; thin to 10–12" Row spacing: 1½–2'	Half-hardy	Full sun to very light shade with some afternoon shade in hot climates. Evenly moist, well-drained soil. Sow seeds outdoors in early spring or start indoors in late winter. Harvest before heads bolt.
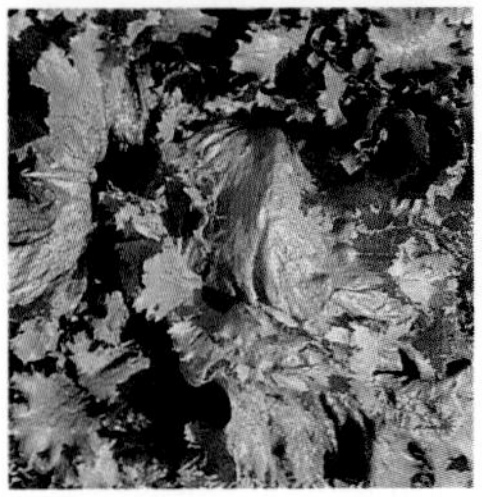	**LETTUCE** CRISPHEAD LETTUCE *Lactuca sativa*	The familiar, tight, round-headed salad green also known as iceberg lettuce. As plants mature, thin seedlings to provide room for the developing heads and use the thinnings as leaf lettuce. Cultivars have medium to light green or red leaves.	6–12"	45–50 days for leaves; 75–110 days for heads	Depth: ¼–½" Spacing: sow 1" apart; thin to 10–12" Row spacing: 1½–2'	Half-hardy	Full sun to very light shade with some afternoon shade in hot climates. Well-drained soil that never completely dries out. Sow seeds outdoors in early spring or start indoors in late winter. Harvest before heads bolt.

Vegetables for American Gardens

		Plant Height	Time to Harvest	Planting Guide	Hardiness	Growing Conditions
LETTUCE LOOSELEAF LETTUCE *Lactuca sativa*	Easily grown, leafy hybrid and cultivar lettuces that do not form a head. Colors range from green to red, and leaves range from broad and flat to deeply lobed and curly. Harvest outer leaves, leaving the plant center to grow new ones.	6–12"	50–80 days	Depth: ¼–½" Spacing: 6–8" Row spacing: 1–1½'	Half-hardy	Full sun to very light shade with some afternoon shade in warm climates. Well-drained soil that never completely dries out. Sow seeds outdoors from early spring to late summer at 2-week intervals. Harvest before leaves become bitter.
MALABAR SPINACH SUMMER SPINACH *Basella alba*	A perennial vine grown as an annual and a good substitute for garden spinach during the heat of summer. Plants produce succulent, strong-flavored stems and leaves. Red and green cultivars are available.	10–12'	90–120 days	Depth: ¼" Spacing: 16–20" Row spacing: 2–2½'	Tender	Full sun to light shade. Fertile, well-drained soil. Plants respond well to high nitrogen levels. Start seeds indoors in late winter in zone 6 and colder; otherwise sow seeds outdoors when danger of frost has passed. Harvest smaller leaves.
MELON CANTALOUPE *Cucumis melo* Cantalupensis group MUSKMELON *C. melo* Reticulatus group	Annual vines grown for their 4- to 7-in. ribbed fruits whose tan or green skins have a netted pattern. The flesh is usually orange, but cultivars are available with white or pink flesh. Harvest when stem slips from fruit with gentle pressure.	1–2'	65–100 days	Depth: ½" Spacing: 2–8' Row spacing: 5–7'	Tender	Full sun. Well-drained soil that is rich in humus. Start seeds in peat pots 3 weeks before last frost or sow outdoors after all danger of frost has passed. Seedlings need warm soil to grow and set fruit. Plastic mulch may help in cool climates.
MELON CRENSHAW MELON *Cucumis melo* Inodorus group	Melons with dark green, tender rinds that turn yellow when the 6- to 12-in. fruits are ripe. The flesh is salmon pink or light green. Crenshaws ripen toward autumn. Harvest fruit by pruning stem with a knife rather than pulling it off the vine.	1–2'	90–100 days	Depth: ½" Spacing: 2–8' Row spacing: 5–7'	Tender	Full sun. Well-drained soil that is rich in humus. Start seeds in peat pots 3 weeks before last frost or sow outdoors after all danger of frost has passed. Seedlings need warm soil to grow and set fruit. Plastic mulch may help in cool climates.
MELON HONEYDEW MELON *Cucumis melo* Inodorus group	Melons with light green or white skins that turn pale yellow when the 5- to 9-in. fruits are ripe. The flesh is white or light green. Honeydews ripen toward autumn. Harvest by pruning stem with a knife rather than pulling it off the vine.	1–2'	80–95 days	Depth: ½" Spacing: 2–8' Row spacing: 5–7'	Tender	Full sun. Well-drained soil that is rich in humus. Start seeds in peat pots 3 weeks before last frost or sow outdoors after all danger of frost has passed. Seedlings need warm soil to grow and set fruit. Plastic mulch may help in cool climates.

		Plant Height	Time to Harvest	Planting Guide	Hardiness	Growing Conditions
	MUSHROOM *Agaricus bisporus*	2–3"	40–90 days	Follow directions from supplier	Tender	Low light, preferably deep shade. Humus-rich, evenly moist, cool soil. Kits for growing indoors usually are available from mid-autumn to early spring. Harvest by sharply twisting the stalk.
	A versatile food, actually an edible fungus, usually raised indoors from kits. The white threads that spread through the soil absorbing nutrients produce small (1- to 2-in.), creamy white, buttonlike mushrooms if grown under proper conditions.					
	MUSTARD MUSTARD GREENS *Brassica juncea*	1–1½'	40–50 days	Depth: ¼–½" Spacing: sow 1" apart; thin to 4–6" Row spacing: 1½–2'	Hardy	Full sun to light shade. Average garden soil. Sow seeds outdoors in early spring and at weekly intervals thereafter. For best flavor, grow and harvest in cool weather. Harvest leaves before plants flower. Aphids can be a problem.
	An annual whose leaves have a zesty flavor that makes them a tasty addition to salads or cooked greens. Cultivars with curled or purple leaves and mild to hot taste are available.					
	NASTURTIUM *Tropaeolum majus*	1–10'	45–90 days	Depth: ¼–½" Spacing: 1½–2' Row spacing: 2–3'	Tender	Full sun to partial shade. Average, well-drained soil. For abundant flowers do not add fertilizer. Plants grow best in cool weather. Sow seeds outdoors where plants are desired after all danger of frost has passed.
	An annual grown for its showy flowers and attractive foliage, both of which are eaten in salads and as a garnish. Flowers are yellow or orange, often spotted or streaked with red. Both trailing and bush forms are available.					
	NEW ZEALAND SPINACH *Tetragonia tetragonioides*	1–2'	60–90 days	Depth: ¼–½" Spacing: 8–12" Row spacing: 2–3'	Half-hardy	Full sun to light shade. Well-drained, sandy soil rich in lime but low in nitrogen. Soak seeds overnight in warm water and sow them outdoors in mid-spring. Provide support when seedlings appear. Plants may self-sow and become weedy.
	A sprawling perennial grown as an annual. A midsummer alternative to garden spinach, New Zealand spinach has triangular, 2- to 4-in. leaves with a strong flavor. Harvest small leaves and use them raw or cooked. Old leaves are tough and bitter.					
	OKRA *Abelmoschus esculentus*	3–6'	55–70 days	Depth: ¼" Spacing: 12" Row spacing: 2–3'	Tender	Full sun. Nutrient-rich, clayey-loam soil. Soak seeds to wash off mucilage prior to planting. In zone 7 and colder start indoors 4 to 5 weeks before last frost; elsewhere sow outdoors after frost.
	An upright, bushy annual bearing yellow hollyhock-like flowers with brown centers that produce 4- to 10-in., green or scarlet, ribbed pods. A favorite vegetable in southern cooking. Harvest by pruning stems when pods are 2–3 in. long.					

Vegetables for American Gardens

			Plant Height	Time to Harvest	Planting Guide	Hardiness	Growing Conditions
	ONION BULB ONION, GLOBE ONION Cepa group	A biennial grown as an annual for its pungent bulbs and sometimes for its hollow, green, succulent stems. Produces attractive, round clusters of white, $1/2$-in. flowers. Cultivars vary in hours of daylight needed to form bulbs.	1–4'	75–120 days	Depth: $1/2$" for seeds; 1" for sets Spacing: sow seeds $1/2$" apart; thin to 3–8" Row spacing: 1–$1 1/2$'	Half-hardy	Full sun. Well-drained soil. Avoid soggy conditions. Sow seeds indoors 10–12 weeks before last frost or plant small bulbs ("sets") outdoors around the time of last frost. When tops yellow, break stems over, allow to wither, and dig up bulbs.
	ONION BUNCHING ONION, WELSH ONION Allium fistulosum	A perennial that forms clumps with time. Bunching onions do not form globular bulbs; they produce thickened cylindrical stems. Their dark green leaves can be harvested as scallions, or the entire plant can be dug up.	6–18"	60–70 days	Depth: $1/2$" for seeds; 1" for sets Spacing: sow seeds $1/2$" apart; thin to 2–6" Row spacing: 1'	Half-hardy	Full sun. Fertile, well-drained soil. Raise from seeds sown indoors 10–12 weeks before last frost or from small bulbs ("sets") planted outdoors around the time of last frost. The cut tops will resprout.
	ONION EGYPTIAN ONION, TOP ONION, TREE ONION Allium cepa Proliferum group	A large, hardy plant that forms hot-tasting bulblets at the tops of succulent stems. Young shoots can be used as scallions. Harvest top bulbs from summer through autumn. Bulbs left on the plant will fall to the ground and root for next year's crop.	2–4'	90–110 days	Depth: $1/2$" for seeds; 1" for sets Spacing: sow seeds $1/2$" apart; thin to 8–12" Row spacing: 1–$1 1/2$'	Hardy	Full sun. Fertile, well-drained soil. Raise from small bulbs ("sets") planted in mid-spring. When tops yellow, break stems over and remove bulblets as they dry. Save small bulblets to plant next year or let them root and overwinter outdoors.
	ONION MULTIPLIER ONION, SHALLOT Allium cepa Aggregatum group	A delicate-flavored onion favored in French cuisine. Plants produce no seeds but multiply by forming brown-skinned bulbs. Use tubular leaves as scallions. Harvest bulbs after tops dry out.	1–$2 1/2$'	90–120 days	Depth: $1/2$" for seeds; 1" for sets Spacing: sow seeds $1/2$" apart; thin to 6–8" Row spacing: 1–$1 1/2$'	Hardy	Full sun. Fertile, well-drained soil. Plants are easy to raise from small bulbs ("scts") planted outdoors or in pots in mid-spring. When tops yellow, break stems over, allow to wither, and dig up bulbs. Save small ones for sets.
	PARSLEY Petroselinum crispum var. crispum	A biennial herb grown as an annual and used as a potherb or as a garnish for its deep green, pungent foliage. Varieties differ in amount of leaf curliness. Flat varieties tend to have more pungent flavor.	6–12"	70–80 days	Depth: $1/4$–$1/2$" Spacing: 6–8" Row spacing: 1–$1 1/2$'	Very hardy	Full sun to partial shade. Moist, well-drained soil of average fertility. Soak seeds to hasten germination. Sow seeds indoors 8–10 weeks before last frost. Seeds germinate slowly (3–4 weeks). Transplant outdoors around last frost.

			Plant Height	Time to Harvest	Planting Guide	Hardiness	Growing Conditions
	PARSLEY HAMBURG PARSLEY, ROOT PARSLEY *Petroselinum crispum* var. *tuberosum*	A biennial grown as an annual. The parsniplike root and flat, parsley leaves are both used for seasoning. The creamy white taproot is often grated and used raw or cooked.	1–1½'	80–95 days	Depth: ¼–½" Spacing: 6–12" Row spacing: 1–2'	Very hardy	Full sun to partial shade. Moist, well-drained soil of average fertility. Soak seeds to hasten germination. Sow seeds indoors 8–10 weeks before last frost. Seeds germinate slowly (3–4 weeks). Transplant outdoors after last frost.
	PARSNIP *Pastinaca sativa*	A tall biennial with divided, long, celery-like leaves. It is grown for its edible taproots, which have a sweet but tangy flavor. Harvest the 8- to 15-in. white roots in autumn through early winter and again in early spring. Freezing improves flavor.	3–4'	80–120 days	Depth: ½" Spacing: sow 1" apart; thin to 3" Row spacing: 1½–2'	Very hardy	Full sun. Deep, well-drained, evenly moist, sandy-loam soil. Sow outdoors in early spring. Seeds germinate slowly. Overplant parsnip beds with radish seeds; the quickly emerging radishes loosen soil for slow-germinating parsnips.
	PEA BLACK-EYED PEA, COWPEA, *Vigna unguiculata* subsp. *unguiculata*	An annual legume, sometimes called Southern pea, whose pods grow in clusters and reach 6–9 in. long. Pods are edible when small, but more frequently are harvested when mature and shelled for dry peas. Both vine and bush forms are available.	2–10'	65–90 days	Depth: 1" Spacing: sow 2–4" apart; thin to 6" Row spacing: 2½–4'	Tender	Full sun. Well-drained soil. Plants grow best in hot weather. Sow seeds outdoors 2 weeks after last frost and provide supports for vine cultivars. Harvest before pods are completely dry.
	PEA GARDEN PEA, SHELLING PEA *Pisum sativum* var. *sativum*	An annual vegetable grown for its moist seeds. Most cultivars are tender vines that climb by tendrils, but dwarf, bushy cultivars are also available. The flowers are typically white. Harvest when pods begin to plump.	1½–5'	55–70 days	Depth: ½–1" Spacing: 3–4" Row spacing: 2–3'	Hardy	Full sun to partial shade. Well-drained, moist soil. Do not provide additional nitrogen fertilizer. Peas grow best in cool weather. Sow seeds outdoors in early spring as soon as soil can be worked. Provide supports for tall varieties.
	PEA SNAP PEA *Pisum sativum* var. *sativum*	Edible-pod peas with 2- to 3-in. rounded, rather than flattened, pods. Some cultivars have strings that should be removed before cooking; others are stringless. Harvest when young or just before pods start to yellow.	2–7'	55–70 days	Depth: ½–1" Spacing: 3–4" Row spacing: 3–4'	Hardy	Full sun to partial shade. Well-drained, moist soil. Do not provide additional nitrogen fertilizer. Peas grow best in cool weather. Sow seeds outdoors in early spring as soon as soil can be worked. Provide supports for tall varieties.

Vegetables for American Gardens

			Plant Height	Time to Harvest	Planting Guide	Hardiness	Growing Conditions
PEA SNOW PEA, SUGAR PEA *Pisum sativum* var. *macrocarpon*		*The classic edible-pod pea with 2- to 4-in. flattened pods that are harvested just as the seeds inside start to bulge. Small pods are more tender and do not need to have strings removed before cooking.*	1–5'	60–70 days	Depth: 1/2–1" Spacing: 3–4" Row spacing: 2–3'	Half-hardy	*Full sun to partial shade. Well-drained, moist soil. Do not provide additional nitrogen fertilizer. Peas grow best in cool weather. Sow seeds outdoors in early spring as soon as soil can be worked. Provide supports for tall varieties.*
PEANUT *Arachis hypogaea*		*An annual related to peas and beans. Yellow flowers grow into "pegs" that point downward and grow into the soil, forming pod fruits underground. Running and bush cultivars are available. Peanuts require a long, warm growing season.*	1–1½'	120–150 days	Depth: 1–1½" Spacing: 1' Row spacing: 2½–3'	Tender	*Full sun. Sandy, well-drained soil that is evenly moist. In zones 6 and colder, start seeds indoors in peat pots 4 weeks before last frost. Elsewhere sow seeds outdoors after last frost. Mound soil around plants as pegs appear.*
PEPPER BELL PEPPER, SWEET PEPPER *Capsicum annuum* Grossum group		*A frost-sensitive native perennial bearing the familiar sweet fruits used in salads and cooking. The blocky bell and elongated banana are two forms of sweet pepper. Cultivars range in color from green to yellow, red, brown, orange, or lavender.*	2–3'	50–75 days	Depth: 1/4" Spacing: 1–2' Row spacing: 2½–3'	Tender	*Full sun. Well-drained, evenly moist, fertile soil. Plants grow best in warm weather. Start seeds indoors 8–10 weeks before last frost and transplant after all danger of frost has passed. Stocky seedlings are best for transplanting.*
PEPPER CHILI PEPPER, HOT PEPPER *Capsicum annuum* Longum group		*A frost-sensitive perennial whose many cultivars range from relatively mild 'Anaheim' to medium-hot 'Jalapeño', to the scorching 'Habañero'. Usually chilies turn from green to red or yellow, and pungency increases with ripening.*	2–3'	60–80 days	Depth: 1/4" Spacing: 1–2' Row spacing: 2½–3'	Tender	*Full sun. Well-drained, evenly moist, fertile soil. Plants grow best in warm weather. Start seeds indoors 8–10 weeks before last frost and transplant after all danger of frost has passed. Stocky seedlings are best for transplanting.*
PEPPER PIMIENTO PEPPER *Capsicum annuum* Grossum group 'Pimiento'		*A frost-sensitive perennial native to the American tropics. This very sweet cultivar of bell pepper has red fruits used in salads or marinated for stuffing green olives.*	2–3'	50–75 days	Depth: 1/4" Spacing: 1–2' Row spacing: 2½–3'	Tender	*Full sun. Well-drained, evenly moist, fertile soil. Plants grow best in warm weather. Start seeds indoors 8–10 weeks before last frost and transplant after all danger of frost has passed. Stocky seedlings are best for transplanting.*

			Plant Height	Time to Harvest	Planting Guide	Hardiness	Growing Conditions
	POTATO *Solanum tuberosum*	*A perennial grown as an annual for its edible tubers. Cultivars vary in size, skin color, flesh color, and texture. For spring harvest dig up small tubers when plants first blossom; for autumn harvest dig up mature tubers after first heavy frost.*	1–2½'	*90–120 days*	Depth: 3–4" Spacing: 10–12" Row spacing: 2–3'	Half-hardy	*Full sun. Well-drained, sandy soil rich in humus. At last frost plant small "seed potatoes" or pieces of larger ones. Each planted piece must have 2–3 eyes. Mound soil around growing plants to protect tubers.*
	PUMPKIN *Cucurbita pepo var. pepo*	*A favorite autumn annual vine bearing orange fruits and large, bristly leaves. Fruit sizes range from small "sugar" types used in pies to the mammoth "jack-o'-lantern" types; some have edible seeds lacking hulls.*	1–2½'	*90–115 days*	Depth: ¾–1" Spacing: 1½–2' Row spacing: 6–8'	Tender	*Full sun. Fertile, well-drained soil that is evenly moist. Sow seeds outdoors after all danger of frost has passed. Harvest before frost by pruning fruit with clippers or a knife, leaving a 3-in. stub attached to pumpkin.*
	PURSLANE KITCHEN-GARDEN PURSLANE *Portulaca oleracea var. sativa*	*A sprawling annual whose succulent red or orange stems and 2-in., spatula-shaped green leaves are used as a crunchy addition to mixed salads or cooked as greens. This can be grown as a container plant.*	3"	*45–60 days*	Depth: ⅛–¼" Spacing: sow ½" apart; thin to 4–6" Row spacing: 8–12"	Tender	*Full sun. Well-drained, average garden soil. Plants tolerate sandy soil. Sow seeds outdoors after last frost. Plants resprout after harvest; they can also be propagated from stem pieces. Plants spread rapidly and may become weedy.*
	RADICCHIO RED CHICORY *Cichorium intybus*	*A perennial that produces red, rounded, somewhat bitter leaves in clustered heads that look like a small red cabbage. Harvest when head turns red or pull roots in autumn and force like Belgian endive. (See CHICORY.)*	6–10"	*55–110 days*	Depth: ¼–½" Spacing: 9–14" Row spacing: 1–2'	Hardy	*Full sun to light shade. Well-drained, fertile, humus-rich soil. Keep the soil evenly moist to prevent leaf wilting. Sow seeds outdoors in mid-spring, even before last frost in zone 6 and colder. For an autumn crop elsewhere, sow outdoors in midsummer.*
	RADISH *Raphanus sativus*	*An easy, fast-growing vegetable grown for its crisp, fleshy, pungent taproot. Rough leaves grow in a clumped rosette. Harvest before the flower-bearing shoot bolts and displays yellow, 4-petaled flowers.*	6–12"	*27–35 days*	Depth: ½" Spacing: 1–2" Row spacing: 1'	Half-hardy	*Full sun. Moist, well-drained soil of moderate fertility. Plants grow best in cool weather. Sow seeds outdoors at frequent intervals starting in early spring as soon as ground can be worked. Harvest as soon as taproots attain full size.*

Vegetables for American Gardens

			Plant Height	Time to Harvest	Planting Guide	Hardiness	Growing Conditions
	RADISH DAIKON, WINTER RADISH *Raphanus sativus* 'Longipinnatus'	*An Asian radish larger than the common table radish. The flesh of these carrot-sized radishes is white. It is used raw, pickled, or cooked, especially in stir-fry dishes.*	6–12"	40–70 days	Depth: 1/2" Spacing: sow 2" apart; thin to 4–6" Row spacing: 18"	Hardy	*Full sun. Moist, well-drained soil of moderate fertility. Plants grow best in cool weather. For summer harvest sow seeds outdoors in early spring as soon as soil can be worked. Sow seeds again in late summer to harvest in late autumn into winter.*
	RHUBARB PIE PLANT *Rheum rhabarbarum*	*A perennial with woody rootstocks and broad, large (1- to 2-ft.), red-stalked leaves. Only the stalks are edible; discard the poisonous green leaf blade. Harvest by gently pulling stalks. Cut off flower buds to prolong harvest.*	2–3'	2nd year from spring to mid-summer	Depth: 2" Spacing: 2 1/2–3' Row spacing: 3'	Hardy	*Full sun to light shade. Humus-rich, well-drained, evenly moist soil. Propagate from root cuttings planted in spring. Do not harvest for at least a year, and preferably two. Slugs and snails can be a problem.*
	RUTABAGA *Brassica napus* Napobrassica group	*A giant turnip with a bulbous white- or yellow-fleshed taproot weighing up to 7 lbs. The blue-green leaves are edible as well. Coating the root with paraffin wax will extend its shelf life.*	1–1 1/2'	90–120 days	Depth: 1/2" Spacing: 6–9" Row spacing: 1 1/2–2'	Very hardy	*Full sun. Well-drained, compost-rich soil. Grow as a fall crop. Sow seeds outdoors in late spring or summer, about 15 weeks before first frost. Harvest the entire plant in autumn to early winter.*
	SALSIFY OYSTER PLANT *Tragopogon porrifolius*	*A hardy biennial whose white, 10- to 15-in. taproot has the flavor of oysters. The leaves of the large, dandelion-like plant are narrow, and the flowers are either yellow or purple.*	2–3'	110–130 days	Depth: 1/4" Spacing: sow 2" apart; thin to 4–6" Row spacing: 1–1 1/2'	Hardy	*Full sun. Well-drained, sandy soil that never dries out completely. In zone 7 and cooler sow outdoors in early spring and harvest in autumn. Elsewhere sow in autumn for winter through early spring harvest.*
	SCORZONERA BLACK OYSTER PLANT, BLACK SALSIFY *Scorzonera hispanica*	*A perennial grown for its long, fleshy, oyster-flavored taproot. This plant resembles salsify but with broader leaves and shorter taproots that have black skin and white flesh. Carefully harvest the roots, as they are fragile.*	2–3'	110–130 days	Depth: 1/4" Spacing: sow 2" apart; thin to 4–6" Row spacing: 1–1 1/2'	Hardy	*Full sun. Well-drained, sandy soil that never dries out completely. Sow outdoors in early spring and harvest in late autumn or early winter. Cool weather improves flavor.*

			Plant Height	Time to Harvest	Planting Guide	Hardiness	Growing Conditions
SORREL FRENCH SORREL, GARDEN SORREL *Rumex acetosa*		*A perennial herb whose tart-flavored leaves are used as a seasoning in soups and salads. Most of the lance-shaped leaves grow near the base of the plant. Harvest young leaves as needed.*	2–3'	60–90 days	Depth: 1/4" Spacing: 10–12" Row spacing: 1–1 1/2'	Hardy	*Full sun to light shade. Average garden soil. Sow seeds outdoors in early spring. To prolong harvest, do not allow plants to flower. To encourage the growth of new leaves, periodically cut stems back to near the ground.*
SOYBEAN *Glycine max*		*A bushy annual legume high in protein and oil and suitable for the home garden. Choose home garden cultivars such as 'Prize' and 'Okuhara'. Harvest the hairy pods when they start to yellow. Remove beans by boiling pods.*	2–3'	60–115 days	Depth: 1 1/2–2" Spacing: 5–9" Row spacing: 2–3'	Tender	*Full sun. Fertile, moist, well-drained soil. Do not apply additional nitrogen fertilizer. Plants grow best in warm weather. Sow seeds outdoors after last frost.*
SPINACH *Spinacia oleracea*		*A cool-season annual whose deep green, long-stemmed leaves are used raw in salads or cooked as greens. Harvest leaves individually as needed before plants flower.*	6–12"	45–55 days	Depth: 1/2" Spacing: 6–9" Row spacing: 1–1 1/2'	Hardy	*Full sun to light shade. Well-drained, fertile, evenly moist soil. Sow seeds outdoors in very early spring. Sow again every few weeks until late spring. In warmer zones sow outdoors in autumn, over-winter under mulch, and harvest in spring.*
SQUASH, SUMMER COURGETTE, ZUCCHINI *Cucurbita pepo* var. *melopepo*		*A popular vining annual with smooth green club-shaped or curved fruits. Harvest when fruits are less than 8 in. long and are still tender. Bush cultivars and yellow-fruiting cultivars are available.*	1–3'	50–55 days	Depth: 1/2–1" Spacing: 2–2 1/2' Row spacing: 3–5'	Tender	*Full sun. Fertile, moist, well-drained soil. Plants grow best in warm weather. In zones 5 and colder start plants indoors in peat pots. Elsewhere sow seeds outdoors when all danger of frost has passed. Provide support for vining cultivars.*
SQUASH, SUMMER PATTYPAN SQUASH, SCALLOP SQUASH *Cucurbita pepo* var. *melopepo*		*A vining annual bearing 3- to 4-in., discus-shaped fruits with scalloped knobby edges. The skin may be solid or striped, white, green, or yellow. Harvest while fruits are small and tender.*	1–3'	50–55 days	Depth: 1/2–1" Spacing: 2–2 1/2" Row spacing: 3–5'	Tender	*Full sun. Fertile, moist, well-drained soil. Plants grow best in warm weather. In zones 5 and colder start plants indoors in peat pots. Elsewhere sow seeds outdoors when all danger of frost has passed. Provide support for vining cultivars.*

Vegetables for American Gardens

		Plant Height	Time to Harvest	Planting Guide	Hardiness	Growing Conditions
SQUASH, SUMMER YELLOW SUMMER SQUASH *Cucurbita pepo* *var. melopepo*	A vining annual bearing crook-necked, elongated yellow fruits, sometimes with a warty surface and usually with fine, bristly hairs. Harvest small fruits (4–6 in.) when the seeds and skin are tender. Bush cultivars are available.	1–3'	50–55 days	Depth: 1/2–1" Spacing: 2-2½' Row spacing: 3–5'	Tender	Full sun. Fertile, moist, well-drained soil. Plants grow best in warm weather. In zones 5 and colder start plants indoors in peat pots. Elsewhere sow seeds outdoors when all danger of frost has passed. Provide support for vining cultivars.
SQUASH, WINTER ACORN SQUASH *Cucurbita pepo* *var. pepo*	A vining annual bearing fluted 4- to 6-in. fruits with dark green or golden skin and pale yellow flesh. Harvest before frost when skin is hard; prune fruit from vine leaving a 3-in. stem attached to the squash. Bush cultivars are available.	1–1¼'	80–100 days	Depth: 3/4–1" Spacing: 1½–2' Row spacing: 3–5' for bush types; 6–8' for vines	Tender	Full sun. Fertile, moist, well-drained soil. Plants grow best in warm weather. Sow seeds outdoors when all danger of frost has passed. At the end of summer pinch off growing tips to stop flowering and promote ripening of existing fruit.
SQUASH, WINTER BUTTERCUP SQUASH, TURBAN SQUASH *Cucurbita maxima*	A vining annual bearing odd-shaped squash that look like small (5- to 6-in.) flattened pumpkins with a ring-like disk at the blossom end. The orange flesh is smooth and sweet. Harvest when skin is hard, leaving a 3-in. stem attached to the squash.	1–2'	90–100 days	Depth: 3/4–1" Spacing: 2–3' Row spacing: 6–8'	Tender	Full sun. Fertile, moist, well-drained soil. Plants grow best in warm weather. Sow seeds outdoors when all danger of frost has passed. At the end of summer pinch off growing tips to stop flowering and promote ripening of existing fruit
SQUASH, WINTER BUTTERNUT SQUASH *Cucurbita moschata*	An annual vine that bears large (6- to 12-in.), light tan or orange fruits that look like elongated bells. The smooth, sweet, orange flesh nearly fills the entire fruit. Harvest when skin is hard, leaving a 3-in. stem attached to the squash.	1–2'	75–85 days	Depth: 3/4–1" Spacing: 2–3' Row spacing: 6–8'	Tender	Full sun. Fertile, moist, well-drained soil. Plants grow best in warm weather. Sow seeds outdoors when all danger of frost has passed. At the end of summer pinch off growing tips to stop flowering and promote ripening of existing fruit.
SQUASH, WINTER HUBBARD SQUASH *Cucurbita maxima*	An annual vine bearing torpedo-shaped squash—one of the largest, with 8- to 20-in. fruits. The hard, mottled or warty skin is usually light blue-green, and the flesh is yellow. Harvest by pruning, leaving a stem attached to the squash.	1–2'	90–120 days	Depth: 3/4–1" Spacing: 2–3' Row spacing: 6–8'	Tender	Full sun. Fertile, moist, well-drained soil. Plants grow best in warm weather. Sow seeds outdoors when all danger of frost has passed. At the end of summer pinch off growing tips to stop flowering and promote ripening of existing fruit.

			Plant Height	Time to Harvest	Planting Guide	Hardiness	Growing Conditions
	SQUASH, WINTER SPAGHETTI SQUASH, VEGETABLE SPAGHETTI *Cucurbita pepo*	An annual vine bearing 6- to 9-in., ovoid fruits whose flesh breaks apart when it is cooked, forming strands remarkably similar to crunchy pasta. Harvest by pruning, leaving a 3-in. stem attached to the squash.	1–2'	70–90 days	Depth: $3/4$–1" Spacing: 2–3' Row spacing: 6–8'	Tender	Full sun. Fertile, moist, well-drained soil. Plants grow best in warm weather. Sow seeds outdoors when all danger of frost has passed. At the end of summer pinch off growing tips to stop flowering and promote ripening of existing fruit.
	SUNFLOWER *Helianthus annuus*	A North American native annual grown for its large flowers and edible seeds. To harvest, wait until seeds darken or flower heads turn downward. Cut heads with a 2-in. stem; hang upside down. Rub dry heads with a wire brush to extract seeds.	5–10'	70–90 days	Depth: $3/4$–1" Spacing: 2–3' Row spacing: 6–8'	Hardy	Full sun to very light shade. Well-drained, average to dry soil. In early spring sow seeds outdoors in the desired location. This is an aggressively growing plant that requires much room.
	SWEET POTATO *Ipomoea batatas*	A tropical perennial grown as an annual for its tuberous edible root. Both bush and vine cultivars are available. To harvest the tubers dig up before frost with a spading fork, taking care to avoid nicks. Cure tubers in a warm place for a week.	12–15"	150–175 days	Depth: $1/2$–1" Spacing: 1' Row spacing: $2^{1}/2$–3'	Tender	Full sun. Sandy-loam soil. Plants grow best in hot weather. Plant rooted tuber cuttings ("slips") 3–4 weeks after last frost. Bury slips up to the top leaves. In zones 6 and colder, cover soil with black plastic to promote growth.
	TOMATILLO *Physalis ixocarpa*	An annual vine bearing small piquant fruits frequently used in salsa and Mexican dishes. The fruits look like green-yellow, $1^{1}/2$- to 2-in. tomatoes in a green, papery husk. Harvest when husk splits.	3–4'	90–110 days	Depth: 4–6" Spacing: 1–$1^{1}/2$' Row spacing: 3–4'	Tender	Full sun. Fertile, evenly moist, well-drained soil. Plants grow best in warm weather. Start seeds indoors 8 weeks before last frost and transplant after all danger of frost has passed.
	TOMATO *Lycopersicon esculentum*	A frost-sensitive perennial grown as an annual for its versatile fruit. Many different cultivars offer fruits of various sizes and color. Indeterminate cultivars grow until frost and require support. Determinate ("bush") cultivars are also available.	4–10'	55–95 days	Depth: $1/2$" Spacing: 2–3' Row spacing: 3–6'	Tender	Full sun. Fertile, evenly moist, well-drained soil. Plants grow best in warm weather. Start seeds indoors 8 weeks before last frost, and transplant sturdy seedlings after all danger of frost has passed. Provide support for vining cultivars.

Vegetables for American Gardens

		Plant Height	Time to Harvest	Planting Guide	Hardiness	Growing Conditions
TOMATO CHERRY TOMATO *Lycopersicon esculentum* var. *cerasiforme*	A frost-sensitive perennial grown as an annual and bearing hundreds of small (1-in.), juicy, red or yellow fruits. Both determinate ("bush") and dwarf cultivars are available; these are attractive plants for containers and hanging baskets.	2–4'	50–75 days	Depth: $1/2$" Spacing: 2–3' Row spacing: $2^1/2$–3'	Tender	Full sun. Fertile, evenly moist, well-drained soil. Plants grow best in warm weather. Start seeds indoors 8 weeks before last frost, and transplant sturdy seedlings after all danger of frost has passed. Provide support for vining cultivars.
TOMATO PASTE TOMATO, PLUM TOMATO *Lycopersicon esculentum* var. *pyriforme*	A frost-senstive perennial grown as an annual and bearing pear- or plum-shaped fruits that are used more for processing and sauces than fresh. The taller indeterminate cultivars continue to grow until frost and require support.	4–10"	70–80 days	Depth: $1/2$" Spacing: 2–3' Row spacing: 3–4'	Tender	Full sun. Fertile, evenly moist, well-drained soil. Plants grow best in warm weather. Start seeds indoors 8 weeks before last frost, and transplant sturdy seedlings after all danger of frost has passed. Provide support for vining cultivars.
TURNIP *Brassica rapa* Rapifera group	A versatile plant with edible young leaves and tasty, bulbous, white, 2- to 3-in. taproots. Greens are best when 3 to 4 in. long. To harvest roots dig with a spading fork as needed when they are less than 3 in. long.	9–12"	40–45 days for greens; 65–70 days for roots	Depth: $1/2$" Spacing: 2–3' Row spacing: 3–6'	Hardy	Full sun. Well-drained, evenly moist soil with added compost and lime. Plants grow best in cool weather. For a spring crop sow seeds outdoors 3 weeks before last frost. For an autumn crop sow outdoors 8–10 weeks before first frost.
WATERCRESS *Nasturtium officinale*	An aquatic perennial that forms mats on the surface of slow-moving streams. It also can be grown as an annual under normal garden conditions or even in a pot. Harvest the leaves and small stems with scissors.	6–9"	50–60 days	Depth: $1/4$" Spacing: sow seeds 1" apart; thin to 2–3" Row spacing: 1'	Hardy	Light shade. Wet to moist soil. Plants prefer soil rich in lime. Sow seeds outdoors or in pots in early spring. Harvest as needed.
WATERMELON *Citrullus lanatus*	An annual vine bearing sweet, fleshy fruits covered with a shiny hard rind. Cultivars vary in fruit size (6 to 24 in.) and flesh color (red, yellow, or white). Seedless cultivars are also available. Harvest when fruit sounds hollow when tapped.	1–2'	75–100 days	Depth: $1/4$" Spacing: sow $1/2$" apart, thin to 4" Row spacing: 6–12"	Tender	Full sun. Evenly moist, well-drained, fertile soil. Sow seeds outdoors 3 weeks after last frost or indoors in peat pots 3–4 weeks before last frost and transplant 6 weeks later. Plants need a warm soil; in zones 6 and colder cover soil with black plastic.

Plant Hardiness Zone Map

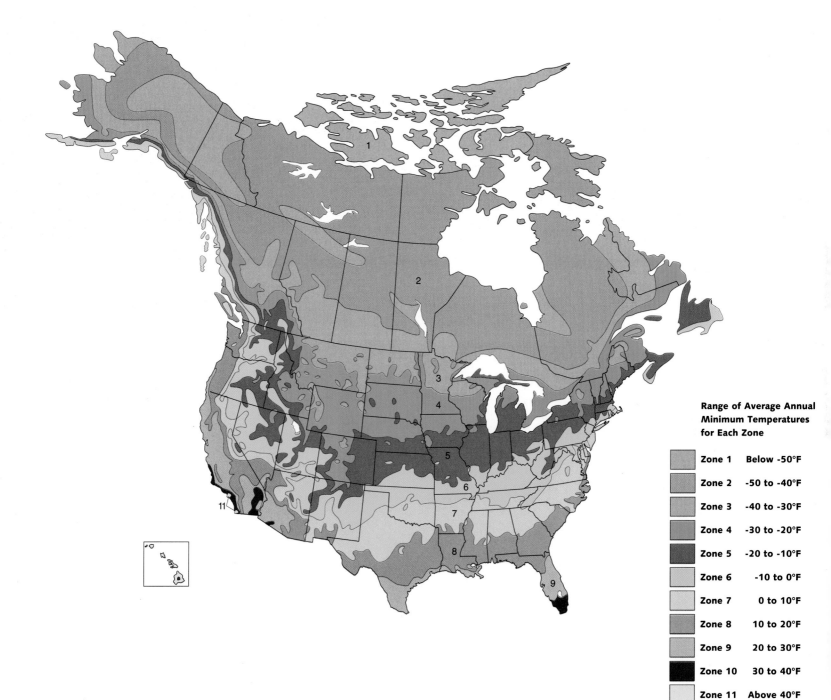

Range of Average Annual Minimum Temperatures for Each Zone

	Zone 1	Below -50°F
	Zone 2	-50 to -40°F
	Zone 3	-40 to -30°F
	Zone 4	-30 to -20°F
	Zone 5	-20 to -10°F
	Zone 6	-10 to 0°F
	Zone 7	0 to 10°F
	Zone 8	10 to 20°F
	Zone 9	20 to 30°F
	Zone 10	30 to 40°F
	Zone 11	Above 40°F

Resources for Vegetable Gardening

There are many dependable mailorder suppliers that can be helpful for vegetable gardeners. A selection is included here. Most have catalogues available upon request (some charge a fee). For further information and supplier suggestions, The Complete Guide to Gardening by Mail is available from the Mailorder Association of Nurseries, Department SCI, 8683 Doves Fly Way, Laurel, MD 20783. Please add $1.00 for postage and handling in the United States ($1.50 for Canada).

Plants and Seeds

Abundant Life Seed Foundation
P.O. Box 772
Port Townsend,
WA 98368
206-385-7192
Nonprofit foundation selling over 600 varieties of open-pollinated, chemical-free seeds.

W. Atlee Burpee Co.
300 Park Avenue
Warminster, PA 18974
215-674-4900
Seeds and supplies from one of the oldest names in American gardening.

Comstock, Ferre & Co.
P.O. Box 125
263 Main Street
Wethersfield, CT 06109
800-753-3773
Vegetable and flower seeds, plants, and supplies.

The Cook's Garden
P.O. Box 535
Londonderry, VT 05148
802-824-3400
Herbs, vegetables, flowers, supplies, and books for gardeners.

Earl May Seed & Nursery
208 N. Elm Street
Shenandoah, IA 51603
712-246-1020
Seeds, supplies, tools, as well as live plants.

Gurney's Seed &
Nursery Co.
110 Capital Street
Yankton, SD 57079
605-665-1930
Offers seeds, plants, and fertilizers for flowers and vegetables.

Harris Seeds
60 Saginaw Drive
P.O. Box 22960
Rochester, NY 14692
716-442-0100
Offers many varieties of vegetable and flower seeds as well as seed-starting equipment.

Henry Field's Seed &
Nursery Co.
415 N. Burnett St.
Shenandoah, IA 51602
605-665-9391
Seeds, supplies, and plants.

Johnny's Selected Seeds
Foss Hill Road
Albion, ME 04910-9731
207-437-4301
Vegetables, herbs, flowers, supplies, and books for gardeners.

J.W. Jung Seed Co.
335 S. High Street
Randolph, WI 53957
800-247-5864
Broad selection of seeds and nursery stock, also tools and supplies.

Mellinger's Inc.
2310 W. South Range Rd
North Lima, OH 44452
800-321-7444
Seeds, plants, supplies, and tools.

Nichols Garden Nursery
1190 N. Pacific Highway
Albany, OR 97321-4598
503-928-9280
Extensive selection of seeds as well as books and supplies.

Park Seed Co.
Cokesbury Road
Greenwood, SC 29647
803-845-3369
Catalogue offers seeds, plants, bulbs, tools, and a wide selection of gardening supplies.

Richters
357 Highway 47
Goodwood, Ontario
Canada LOC 1AO
416-640-6677
Seeds for many herbs and vegetables.

Shepherd's Garden Seeds
6116 Highway 9
Felton, CA 95018
For advice: 408-335-6910
To order: 203-482-3638
Vegetables, herbs, and specialties such as gourmet varieties and selected recipes.

R.H. Shumway Seeds
P.O. Box 1
571 Whaley Pond Road
Graniteville, SC 29829
800-322-7288
Large selection of seeds as well as garden supplies.

Stokes Seeds, Inc.
Box 548
Buffalo, NY 14240-0548
716-695-6980
Vegetable seeds and supplies for commercial farmers and home gardeners.

Thompson & Morgan
P.O. Box 1308
Jackson, NJ 08527-0308
800-274-7333
Seeds of all types and a wide range of other garden supplies.

Tomato Grower's
Supply Co.
P.O. Box 2237
Ft. Myers, FL 33902
813-768-1119
Offers free catalogue specializing in seeds and products for tomato growers.

Regional Specialties

High Altitude Gardens
P.O. Box 1048
Hailey, ID 83333
208-788-4363
Seeds selected for their ability to grow at high altitudes.

Ed Hume Seeds, Inc.
P.O. Box 1450
Kent, WA 98035
206-859-1110
Untreated flower, herb, and vegetable seeds for short-season climates.

Kilgore Seed Co.
1400 W. First Street
Sanford, FL 32771
407-323-6630
Seeds carefully selected for their ability to grow in Florida, Gulf Coast states and other tropical and subtropical areas.

Native Seeds/SEARCH
2509 N. Campbell #325
Tucson, AZ 85719
602-327-9123
Native seeds of the Southwest and Mexico, collected and propagated for preservation purposes; distributed free to Native Americans.

Redwood City Seed Co.
P.O. Box 361
Redwood City, CA 94064
415-325-7333
Heirloom vegetables and herbs, including Asian and Native American varieties.

Seeds Blum
Idaho City Stage
Boise, ID 83706
208-342-0858
Sells vegetables, annuals, and perennials for various conditions.

Southern Exposure Seed Exchange
P.O. Box 158
North Garden, VA 22959
804-973-4703
Catalogue features seeds for southern gardens.

Supplies & Accessories

Alsto's Handy Helpers
P.O. Box 1267
Galesburg, IL 61401
800-447-0048
Easy-to-use tools and gardening equipment.

Country Home Products
P.O. Box 89
Ferry Road
Charlotte, VT 05445
800-446-8746
Mowers, trimmers, clippers, composters, and various garden tools.

Earth-Rite
Zook & Ranck, Inc.
RD 1, Box 243
Gap, PA 17527
800-332-4171
Fertilizers and soil amendments for lawn and garden.

Garden Way, Inc.
102nd St. & 9th Ave.
Troy, NY 12180
800-833-6990
Mowers, rotary tillers, garden carts, and various other lawn and garden equipment.

Gardener's Eden
P.O. Box 7303
San Francisco, CA 94120
800-822-9600
Many items appropriate for gardeners, including outdoor containers, tools, and accessories.

Gardener's Supply Co.
128 Intervale Rd.
Burlington, VT 05401
800-876-5520
Greenhouse kits and materials for building cold frames, as well as a wide selection of other gardening products.

Gardens Alive!
5100 Schenley Place
Lawrenceburg, IN 47025
812-537-8650
Beneficial insects and a complete line of supplies for organic gardening.

Home Gardener
Manufacturing Company
30 Wright Avenue
Lititz, PA 17543
800-880-2345
Offers composting and related home gardening equipment.

Kemp Company
160 Koser Road
Lititz, PA 17543
800-441-5367
Featuring shredders, chippers, and other power equipment.

Plow & Hearth
P.O. Box 830
Orange, VA 22960
800-866-6072
Gardening tools and products as well as garden ornaments and furniture.

Ringer Corporation
9959 Valley View Road
Eden Prairie, MN 55344
612-941-4180
Organic soil amendments, beneficial insects, and garden tools.

Smith & Hawken
Two Arbor Lane
Box 6900
Florence, KY 41022-6900
800-776-3336
Well-crafted tools as well as containers, supplies, and furniture.

Solutions
P.O. Box 6878
Portland, OR 97228
800-342-9988
Offers home and gardening products designed to make jobs easier.

Walt Nicke Co.
P.O. Box 433
36 McLeod Lane
Topsfield, MA 01983
800-822-4114
Catalogue features over 300 tools and products.

Underglass Mfg. Corp.
P.O. Box 323
Wappingers Falls,
NY 12590
914-298-0645
Color catalogue features greenhouses and solariums. Offers custom-built cold frames.

Index

Photo Credits

All photography credited as follows is copyright © 1994 by the individual photographers.

Karen Bussolini: pp. 12, 25 (center and right), 33, 55 (top left and center), 62, 63, 69 (bottom), 81, 95 (bottom); **David Cavagnaro:** pp. 8, 9 (all), 88 (top right, bottom left and right), 93 (all), 100; **Walter Chandoha:** p. 28; **Rosalind Creasy:** pp. 10, 11 (bottom left and right), 15, 16, 24 (left and center), 25 (left), 26, 27 (all), 30, 31, 53, 55 (top right, bottom left and right), 64, 66, 97 (top left and right); **Ken Druse:** pp. 4, 7, 14, 17 (top left), 19 (top), 36, 60; **Derek Fell:** p. 51; **Dency Kane:** pp. 11 (top left and center), 24 (right), 72 (right), 74, 77, 97 (bottom left); **Robert Kourik:** pp. 17 (bottom left), 82, 86 (center right and far right); **Maggie Oster:** p. 37; **Jerry Pavia:** pp. 19 (bottom left and right), 20, 21, 59, 78; **Joanne Pavia:** pp. 29, 52, 97 (bottom right); **Michael S. Thompson:** pp. 11 (top right), 17 (bottom right), 22, 32, 46, 49 (bottom), 50, 67, 72 (left), 86 (far left and center left), 88 (top left), 89, 91.

Step-by-step photography by Derek Fell.

Front cover photograph copyright © 1994 by Derek Fell.

All plant encyclopedia photography is copyright © 1994 by **Derek Fell**, except the following, which are copyright © 1994 by **Rosalind Creasy:** *Vicia faba, Phaseolus vulgaris, Lepidium sativum, Cucumis melo, Rumex acetosa.*